COMPATIBILITY, STABILITY, AND SHEAVES

PURE AND APPLIED MATHEMATICS

A Program of Monographs, Textbooks, and Lecture Notes

MONOGRAPHS AND TEXTBOOKS IN PURE AND APPLIED MATHEMATICS

1. *K. Yano*, Integral Formulas in Riemannian Geometry (1970)
2. *S. Kobayashi*, Hyperbolic Manifolds and Holomorphic Mappings (1970)
3. *V. S. Vladimirov*, Equations of Mathematical Physics (A. Jeffrey, ed.; A. Littlewood, trans.) (1970)
4. *B. N. Pshenichnyi*, Necessary Conditions for an Extremum (L. Neustadt, translation ed.; K. Makowski, trans.) (1971)
5. *L. Narici et al.*, Functional Analysis and Valuation Theory (1971)
6. *S. S. Passman*, Infinite Group Rings (1971)
7. *L. Dornhoff*, Group Representation Theory. Part A: Ordinary Representation Theory. Part B: Modular Representation Theory (1971, 1972)
8. *W. Boothby and G. L. Weiss, eds.*, Symmetric Spaces (1972)
9. *Y. Matsushima*, Differentiable Manifolds (E. T. Kobayashi, trans.) (1972)
10. *L. E. Ward, Jr.*, Topology (1972)
11. *A. Babakhanian*, Cohomological Methods in Group Theory (1972)
12. *R. Gilmer*, Multiplicative Ideal Theory (1972)
13. *J. Yeh*, Stochastic Processes and the Wiener Integral (1973)
14. *J. Barros-Neto*, Introduction to the Theory of Distributions (1973)
15. *R. Larsen*, Functional Analysis (1973)
16. *K. Yano and S. Ishihara*, Tangent and Cotangent Bundles (1973)
17. *C. Procesi*, Rings with Polynomial Identities (1973)
18. *R. Hermann*, Geometry, Physics, and Systems (1973)
19. *N. R. Wallach*, Harmonic Analysis on Homogeneous Spaces (1973)
20. *J. Dieudonné*, Introduction to the Theory of Formal Groups (1973)
21. *I. Vaisman*, Cohomology and Differential Forms (1973)
22. *B.-Y. Chen*, Geometry of Submanifolds (1973)
23. *M. Marcus*, Finite Dimensional Multilinear Algebra (in two parts) (1973, 1975)
24. *R. Larsen*, Banach Algebras (1973)
25. *R. O. Kujala and A. L. Vitter, eds.*, Value Distribution Theory: Part A; Part B: Deficit and Bezout Estimates by Wilhelm Stoll (1973)
26. *K. B. Stolarsky*, Algebraic Numbers and Diophantine Approximation (1974)
27. *A. R. Magid*, The Separable Galois Theory of Commutative Rings (1974)
28. *B. R. McDonald*, Finite Rings with Identity (1974)
29. *J. Satake*, Linear Algebra (S. Koh et al., trans.) (1975)
30. *J. S. Golan*, Localization of Noncommutative Rings (1975)
31. *G. Klambauer*, Mathematical Analysis (1975)
32. *M. K. Agoston*, Algebraic Topology (1976)
33. *K. R. Goodearl*, Ring Theory (1976)
34. *L. E. Mansfield*, Linear Algebra with Geometric Applications (1976)
35. *N. J. Pullman*, Matrix Theory and Its Applications (1976)
36. *B. R. McDonald*, Geometric Algebra Over Local Rings (1976)
37. *C. W. Groetsch*, Generalized Inverses of Linear Operators (1977)
38. *J. E. Kuczkowski and J. L. Gersting*, Abstract Algebra (1977)
39. *C. O. Christenson and W. L. Voxman*, Aspects of Topology (1977)
40. *M. Nagata*, Field Theory (1977)
41. *R. L. Long*, Algebraic Number Theory (1977)
42. *W. F. Pfeffer*, Integrals and Measures (1977)
43. *R. L. Wheeden and A. Zygmund*, Measure and Integral (1977)
44. *J. H. Curtiss*, Introduction to Functions of a Complex Variable (1978)
45. *K. Hrbacek and T. Jech*, Introduction to Set Theory (1978)
46. *W. S. Massey*, Homology and Cohomology Theory (1978)
47. *M. Marcus*, Introduction to Modern Algebra (1978)
48. *E. C. Young*, Vector and Tensor Analysis (1978)
49. *S. B. Nadler, Jr.*, Hyperspaces of Sets (1978)
50. *S. K. Segal*, Topics in Group Kings (1978)
51. *A. C. M. van Rooij*, Non-Archimedean Functional Analysis (1978)
52. *L. Corwin and R. Szczarba*, Calculus in Vector Spaces (1979)

53. *C. Sadosky*, Interpolation of Operators and Singular Integrals (1979)
54. *J. Cronin*, Differential Equations (1980)
55. *C. W. Groetsch*, Elements of Applicable Functional Analysis (1980)
56. *I. Vaisman*, Foundations of Three-Dimensional Euclidean Geometry (1980)
57. *H. I. Freedan*, Deterministic Mathematical Models in Population Ecology (1980)
58. *S. B. Chae*, Lebesgue Integration (1980)
59. *C. S. Rees et al.*, Theory and Applications of Fourier Analysis (1981)
60. *L. Nachbin*, Introduction to Functional Analysis (R. M. Aron, trans.) (1981)
61. *G. Orzech and M. Orzech*, Plane Algebraic Curves (1981)
62. *R. Johnsonbaugh and W. E. Pfaffenberger*, Foundations of Mathematical Analysis (1981)
63. *W. L. Voxman and R. H. Goetschel*, Advanced Calculus (1981)
64. *L. J. Corwin and R. H. Szcarba*, Multivariable Calculus (1982)
65. *V. I. Istrățescu*, Introduction to Linear Operator Theory (1981)
66. *R. D. Järvinen*, Finite and Infinite Dimensional Linear Spaces (1981)
67. *J. K. Beem and P. E. Ehrlich*, Global Lorentzian Geometry (1981)
68. *D. L. Armacost*, The Structure of Locally Compact Abelian Groups (1981)
69. *J. W. Brewer and M. K. Smith, eds.*, Emily Noether: A Tribute (1981)
70. *K. H. Kim*, Boolean Matrix Theory and Applications (1982)
71. *T. W. Wieting*, The Mathematical Theory of Chromatic Plane Ornaments (1982)
72. *D. B. Gauld*, Differential Topology (1982)
73. *R. L. Faber*, Foundations of Euclidean and Non-Euclidean Geometry (1983)
74. *M. Carmeli*, Statistical Theory and Random Matrices (1983)
75. *J. H. Carruth et al.*, The Theory of Topological Semigroups (1983)
76. *R. L. Faber*, Differential Geometry and Relativity Theory (1983)
77. *S. Barnett*, Polynomials and Linear Control Systems (1983)
78. *G. Karpilovsky*, Commutative Group Algebras (1983)
79. *F. Van Oystaeyen and A. Verschoren*, Relative Invariants of Rings (1983)
80. *I. Vaisman*, A First Course in Differential Geometry (1984)
81. *G. W. Swan*, Applications of Optimal Control Theory in Biomedicine (1984)
82. *T. Petrie and J. D. Randall*, Transformation Groups on Manifolds (1984)
83. *K. Goebel and S. Reich*, Uniform Convexity, Hyperbolic Geometry, and Nonexpansive Mappings (1984)
84. *T. Albu and C. Năstăsescu*, Relative Finiteness in Module Theory (1984)
85. *K. Hrbacek and T. Jech*, Introduction to Set Theory: Second Edition (1984)
86. *F. Van Oystaeyen and A. Verschoren*, Relative Invariants of Rings (1984)
87. *B. R. McDonald*, Linear Algebra Over Commutative Rings (1984)
88. *M. Namba*, Geometry of Projective Algebraic Curves (1984)
89. *G. F. Webb*, Theory of Nonlinear Age-Dependent Population Dynamics (1985)
90. *M. R. Bremner et al.*, Tables of Dominant Weight Multiplicities for Representations of Simple Lie Algebras (1985)
91. *A. E. Fekete*, Real Linear Algebra (1985)
92. *S. B. Chae*, Holomorphy and Calculus in Normed Spaces (1985)
93. *A. J. Jerri*, Introduction to Integral Equations with Applications (1985)
94. *G. Karpilovsky*, Projective Representations of Finite Groups (1985)
95. *L. Narici and E. Beckenstein*, Topological Vector Spaces (1985)
96. *J. Weeks*, The Shape of Space (1985)
97. *P. R. Gribik and K. O. Kortanek*, Extremal Methods of Operations Research (1985)
98. *J.-A. Chao and W. A. Woyczynski, eds.*, Probability Theory and Harmonic Analysis (1986)
99. *G. D. Crown et al.*, Abstract Algebra (1986)
100. *J. H. Carruth et al.*, The Theory of Topological Semigroups, Volume 2 (1986)
101. *R. S. Doran and V. A. Belfi*, Characterizations of C*-Algebras (1986)
102. *M. W. Jeter*, Mathematical Programming (1986)
103. *M. Altman*, A Unified Theory of Nonlinear Operator and Evolution Equations with Applications (1986)
104. *A. Verschoren*, Relative Invariants of Sheaves (1987)
105. *R. A. Usmani*, Applied Linear Algebra (1987)
106. *P. Blass and J. Lang*, Zariski Surfaces and Differential Equations in Characteristic $p >$ 0 (1987)
107. *J. A. Reneke et al.*, Structured Hereditary Systems (1987)

108. *H. Büsemann and B. B. Phadke,* Spaces with Distinguished Geodesics (1987)
109. *R. Harte,* Invertibility and Singularity for Bounded Linear Operators (1988)
110. *G. S. Ladde et al.,* Oscillation Theory of Differential Equations with Deviating Arguments (1987)
111. *L. Dudkin et al.,* Iterative Aggregation Theory (1987)
112. *T. Okubo, Differential Geometry* (1987)
113. *D. L. Stancl and M. L. Stancl,* Real Analysis with Point-Set Topology (1987)
114. *T. C. Gard,* Introduction to Stochastic Differential Equations (1988)
115. *S. S. Abhyankar,* Enumerative Combinatorics of Young Tableaux (1988)
116. *H. Strade and R. Farnsteiner,* Modular Lie Algebras and Their Representations (1988)
117. *J. A. Huckaba,* Commutative Rings with Zero Divisors (1988)
118. *W. D. Wallis,* Combinatorial Designs (1988)
119. *W. Więsław,* Topological Fields (1988)
120. *G. Karpilovsky,* Field Theory (1988)
121. *S. Caenepeel and F. Van Oystaeyen,* Brauer Groups and the Cohomology of Graded Rings (1989)
122. *W. Kozlowski,* Modular Function Spaces (1988)
123. *E. Lowen-Colebunders,* Function Classes of Cauchy Continuous Maps (1989)
124. *M. Pavel,* Fundamentals of Pattern Recognition (1989)
125. *V. Lakshmikantham et al.,* Stability Analysis of Nonlinear Systems (1989)
126. *R. Sivaramakrishnan,* The Classical Theory of Arithmetic Functions (1989)
127. *N. A. Watson,* Parabolic Equations on an Infinite Strip (1989)
128. *K. J. Hastings,* Introduction to the Mathematics of Operations Research (1989)
129. *B. Fine,* Algebraic Theory of the Bianchi Groups (1989)
130. *D. N. Dikranjan et al.,* Topological Groups (1989)
131. *J. C. Morgan II,* Point Set Theory (1990)
132. *P. Biler and A. Witkowski,* Problems in Mathematical Analysis (1990)
133. *H. J. Sussmann,* Nonlinear Controllability and Optimal Control (1990)
134. *J.-P. Florens et al.,* Elements of Bayesian Statistics (1990)
135. *N. Shell,* Topological Fields and Near Valuations (1990)
136. *B. F. Doolin and C. F. Martin,* Introduction to Differential Geometry for Engineers (1990)
137. *S. S. Holland, Jr.,* Applied Analysis by the Hilbert Space Method (1990)
138. *J. Okniński,* Semigroup Algebras (1990)
139. *K. Zhu,* Operator Theory in Function Spaces (1990)
140. *G. B. Price,* An Introduction to Multicomplex Spaces and Functions (1991)
141. *R. B. Darst,* Introduction to Linear Programming (1991)
142. *P. L. Sachdev,* Nonlinear Ordinary Differential Equations and Their Applications (1991)
143. *T. Husain,* Orthogonal Schauder Bases (1991)
144. *J. Foran,* Fundamentals of Real Analysis (1991)
145. *W. C. Brown,* Matrices and Vector Spaces (1991)
146. *M. M. Rao and Z. D. Ren,* Theory of Orlicz Spaces (1991)
147. *J. S. Golan and T. Head,* Modules and the Structures of Rings (1991)
148. *C. Small,* Arithmetic of Finite Fields (1991)
149. *K. Yang,* Complex Algebraic Geometry (1991)
150. *D. G. Hoffman et al.,* Coding Theory (1991)
151. *M. O. González,* Classical Complex Analysis (1992)
152. *M. O. González,* Complex Analysis (1992)
153. *L. W. Baggett,* Functional Analysis (1992)
154. *M. Sniedovich,* Dynamic Programming (1992)
155. *R. P. Agarwal,* Difference Equations and Inequalities (1992)
156. *C. Brezinski,* Biorthogonality and Its Applications to Numerical Analysis (1992)
157. *C. Swartz,* An Introduction to Functional Analysis (1992)
158. *S. B. Nadler, Jr.,* Continuum Theory (1992)
159. *M. A. Al-Gwaiz,* Theory of Distributions (1992)
160. *E. Perry,* Geometry: Axiomatic Developments with Problem Solving (1992)
161. *E. Castillo and M. R. Ruiz-Cobo,* Functional Equations and Modelling in Science and Engineering (1992)
162. *A. J. Jerri,* Integral and Discrete Transforms with Applications and Error Analysis (1992)
163. *A. Charlier et al.,* Tensors and the Clifford Algebra (1992)

164. *P. Biler and T. Nadzieja,* Problems and Examples in Differential Equations (1992)
165. *E. Hansen,* Global Optimization Using Interval Analysis (1992)
166. *S. Guerre-Delabrière,* Classical Sequences in Banach Spaces (1992)
167. *Y. C. Wong,* Introductory Theory of Topological Vector Spaces (1992)
168. *S. H. Kulkarni and B. V. Limaye,* Real Function Algebras (1992)
169. *W. C. Brown,* Matrices Over Commutative Rings (1993)
170. *J. Loustau and M. Dillon,* Linear Geometry with Computer Graphics (1993)
171. *W. V. Petryshyn,* Approximation-Solvability of Nonlinear Functional and Differential Equations (1993)
172. *E. C. Young,* Vector and Tensor Analysis: Second Edition (1993)
173. *T. A. Bick,* Elementary Boundary Value Problems (1993)
174. *M. Pavel,* Fundamentals of Pattern Recognition: Second Edition (1993)
175. *S. A. Albeverio et al.,* Noncommutative Distributions (1993)
176. *W. Fulks,* Complex Variables (1993)
177. *M. M. Rao,* Conditional Measures and Applications (1993)
178. *A. Janicki and A. Weron,* Simulation and Chaotic Behavior of α-Stable Stochastic Processes (1994)
179. *P. Neittaanmäki and D. Tiba,* Optimal Control of Nonlinear Parabolic Systems (1994)
180. *J. Cronin,* Differential Equations: Introduction and Qualitative Theory, Second Edition (1994)
181. *S. Heikkilä and V. Lakshmikantham,* Monotone Iterative Techniques for Discontinuous Nonlinear Differential Equations (1994)
182. *X. Mao,* Exponential Stability of Stochastic Differential Equations (1994)
183. *B. S. Thomson,* Symmetric Properties of Real Functions (1994)
184. *J. E. Rubio,* Optimization and Nonstandard Analysis (1994)
185. *J. L. Bueso, P. Jara, and A. Verschoren,* Compatibility, Stability, and Sheaves (1995)

Additional Volumes in Preparation

COMPATIBILITY, STABILITY, AND SHEAVES

J. L. Bueso
P. Jara
University of Granada
Granada, Spain

A. Verschoren
University of Antwerp
Antwerp, Belgium

Marcel Dekker, Inc.   New York • Basel • Hong Kong

Library of Congress Cataloging-in-Publication Data

Bueso, J. L. (José Luis)
    Compatibility, stability, and sheaves / J. L. Bueso, P. Jara, A. Verschoren.
        p.   cm. — (Monographs and textbooks in pure and applied mathematics; 185)
    Includes bibliographical references and index.
    ISBN 0-8247-9589-X
    1. Localization theory. 2. Rings (Algebra) 3. Sheaf theory. I. Jara, P. (Pascual) II. Verschoren, A. III. Title. IV. Series.
QA247.B84 1995
512'.4—dc20   94-32079
                                                                           CIP

The publisher offers discounts on this book when ordered in bulk quantities. For more information, write to Special Sales/Professional Marketing at the address below.

This book is printed on acid-free paper.

MARCEL DEKKER, INC.
270 Madison Avenue, New York, New York 10016

Current printing (last digit):
10 9 8 7 6 5 4 3 2 1

PRINTED IN THE UNITED STATES OF AMERICA

To the memory of our friend Pere Menal.

Preface

Let us be honest about it: we (the authors) lost lots of time discussing the title of this monograph.

In fact, our main problem was (and is!) that we originally wanted to call it *Compatibility, Stability, and Sheaves: Un Ménage à Trois*. However, whereas a "ménage à trois" frequently bears some negative connotations and usually leads to three losers, here the situation is completely different, as the three protagonists of the title (compatibility, stability, and sheaves) appear to benefit by their active interaction. This is certainly not the first time that such a "ménage" bears mathematical fruit, cf. [7, 26], so we (and the reader) shouldn't bother too much about it, anyway. On the other hand, we want to stress the fact that linking the three concepts in the title is by no means artificial. This might come as a surprise, as there is at first glance no obvious connection between them. Indeed, compatibility (between localization and torsion, say) has been introduced in [84] for better understanding of the behaviour of the ideal structure of a non-commutative ring under abstract localization. Stability, on the other hand, introduced in [27], may be viewed as just another way of looking at the Artin-Rees property, which is rather omnipresent (sometimes in a well-hidden form) in many prominent results within commutative algebra. And sheaves, of course, are really what makes "modern" algebraic geometry, representation theory and logic work, just to mention some parts of their natural habitat.

So, how are these concepts connected?

To answer this question briefly – the rest of this monograph is dedicated to providing a more complete answer – we have to go back to elementary algebraic geometry. The structure sheaf $\mathcal{O}_M$ associated to an R-module M on the spectrum $Spec(R)$ of a (commutative) ring R is usually defined by associating to a basic open subset $X(f) \subseteq Spec(R)$ the module of fractions M_f of M at the multiplicative subset of R generated by $0 \neq f \in R$. If one wants to calculate sections over an arbitrary subset $X(I) \subseteq Spec(R)$ associated to an ideal $I < R$ (assuming R to be noetherian, for simplicity's sake), then one has to use Deligne's formula [39], which asserts that

$$\Gamma(X(I), \mathcal{O}_M) = \varinjlim Hom_R(I^n, M).$$

It is fairly easy to see that the second member of this identity may be interpreted as the localization $Q_I(M)$ of M at I, in the sense of Gabriel [27]. In other words, one may view $Q_I(M)$ as the canonical image of M in the quotient category $R\text{-mod}/ \mathcal{T}_I$, where $\mathcal{T}_I$ is the localizing subcategory consisting of all $M \in R\text{-mod}$ which are torsion at I, i.e., with the property that for any $m \in M$ there exists some positive integer n such that $I^n m = 0$.

Now, Deligne's formula may be generalized to arbitrary subsets of $Spec(R)$, which are closed under generization. These subsets are arbitrary intersections of open subsets and correspond (still assuming R to be a commutative noetherian ring) to radicals in $R\text{-mod}$. More precisely, if σ is a radical in $R\text{-mod}$, i.e., a left exact subfunctor of the identity with the property that $\sigma(M/\sigma M) = 0$ for any $M \in R\text{-mod}$, then one may associate to it the subset $\mathcal{K}(\sigma)$ of $Spec(R)$ consisting of all prime ideals $\mathfrak{p}$ of R with the property that $\sigma(R/\mathfrak{p}) = 0$. These subsets are closed under generization and conversely, every subset of $Spec(R)$ which is closed under generization is of this type. For the special case $\sigma = \sigma_I$ (defined by letting $\sigma_I M$ consist of all $m \in M$ with $I^n m = 0$ for some positive integer n), we recover the open subset $X(I)$ as $\mathcal{K}(\sigma_I)$. Let $\mathcal{T}_\sigma \subseteq R\text{-}\mathbf{mod}$ have the property that $\sigma M = M$ (the σ-torsion R-modules); then $R\text{-}\mathbf{mod}/$ $\mathcal{T}_\sigma$ embeds into $R\text{-}\mathbf{mod}$ and we denote by Q_σ the composition $R\text{-}\mathbf{mod} \to$ $R\text{-}\mathbf{mod}/\mathcal{T}_\sigma \hookrightarrow R\text{-}\mathbf{mod}$. It has been proved in [88, 89] that for any $M \in$ $R\text{-}\mathbf{mod}$ we have

$$(*) \quad \Gamma(\mathcal{K}(\sigma), \mathcal{O}_M|\mathcal{K}(\sigma)) = Q_\sigma(M),$$

which generalizes Deligne's formula.

On the other hand, for a noncommutative noetherian ring R, several sheaf constructions over $Spec(R)$ have been considered in the past, all of them adapted to some particular algebraic or representation theoretic problem. Although (some of) these constructions work reasonably well (see the literature list at the end), in general, the analogue of the identity (*) is no longer valid, even for *open* subsets. A careful examination of its different proofs in the commutative case shows that they all depend at some point on the fact that the functors Q_σ and Q_τ commute for any pair of radicals σ and τ in $R\text{-}\mathbf{mod}$, a fact which is no longer valid over noncommutative rings.

Now, again in the commutative case, the first step toward proving that $Q_\sigma Q_\tau = Q_\tau Q_\sigma$ is to show that $\sigma Q_\tau = Q_\tau \sigma$ and similarly, with the roles of σ and τ interchanged – and even this fails in the noncommutative case.

It is exactly here that compatibility enters into the picture, for the identity $\sigma Q_\tau = Q_\tau \sigma$ is one of the alternative definitions of compatibility between σ and τ introduced in [84] and studied further in [90]. Moreover, it has been proved in [61] that two radicals σ and τ are compatible if and only if the sequence of functors

$$0 \to Q_{\sigma \wedge \tau} \to Q_\sigma \oplus Q_\tau \to Q_{\sigma \vee \tau}$$

(where $\sigma \wedge \tau$ resp. $\sigma \vee \tau$ are the meet and join of σ and τ, to be defined below) is exact. The reader familiar with sheaf theory and abstract localization will, of course, realize that the exactness of the above sequence is equivalent to the sheaf axiom over a union of two open sets, at least if open sets are the form $\mathcal{K}(\sigma)$ for some radical σ in R-**mod** and if for $M \in R$-**mod** we associate to $\mathcal{K}(\sigma)$ the localization $Q_\sigma(M)$ of M at σ.

As a special case, let us work over a left noetherian ring R and let us endow $Spec(R)$ with the Zariski topology, its open sets thus being the $X(I) \subseteq Spec(R)$ consisting of all $P \in Spec(R)$ with $I \not\subset P$. For any $M \in R$-**mod**, we may endow $Spec(R)$ with an associated presheaf $\mathcal{O}_M$ by mapping any open subset $X(I) \subseteq Spec(R)$ to the localization $Q_I(M)$ of M at σ_I. Moreover, since all open subsets are now quasicompact, one may reduce the verification of the sheaf axiom to the case of finite coverings and even to coverings by two open subsets.

It is thus clear, in view of the previous discussion, that if we want to construct a structure *sheaf* on $Spec(R)$ with decent properties, then we will have to endow $Spec(R)$ with a subtopology of the Zariski topology, which yields back the whole Zariski topology in the commutative case and such that for all open subsets $X(I)$ in this topology, the corresponding radicals σ_I are mutually compatible in the sense indicated above.

As we will see in this text, all of this is perfectly possible but, unfortunately, not sufficient to realize our aims. Indeed, compatibility of σ and τ alone does not ensure that Q_σ and Q_τ commute, except in special situations, such as the commutative case, for example. On the other hand, we would also like to put some direct assumption on the individual two-sided ideals I of R guaranteeing that the corresponding radicals σ_I are mutually compatible.

So, here is where we need stability. Indeed, it has been proved in [90] that if σ and τ are stable radicals, then they are compatible and, moreover, the associated localization functors Q_σ and Q_τ commute. In particular, it thus finally follows that we may construct the type of structure sheaves we want, by endowing $Spec(R)$ with a topology $T_s(R)$ whose open subsets $X(I)$ yield radicals σ_I in R–mod which are stable. This obviously imposes some restrictions on the rings for which this construction yields reasonable results, as it applies only to rings with a sufficiently large number of ideals I such that σ_I is stable. However, the rings we aim at (enveloping algebras, group rings, pi rings, ...) are all of this type, as one easily verifies taking into account that the stability of σ_I is equivalent to I satisfying the Artin-Rees property.

The previous discussion might have given the reader the impression that everything now works perfectly well – and this was certainly not our intention. Indeed, the above ideas essentially lead to associating to a reasonably general ring R a geometric object (an "affine scheme") $(Spec(R), \mathcal{O}_R)$, where $\mathcal{O}_R$ is the above structure sheaf on $Spec(R)$ endowed with the Zariski topology. However, this construction is useful only if it behaves functorially and this fails in general. Indeed, an arbitrary ring homomorphism $\varphi : R \to S$ does not even induce a morphism $Spec(S) \to Spec(R)$, as the inverse image $\varphi^{-1}(Q)$ of $Q \in Spec(R)$ is not necessarily prime in R. Now, this is again not a very serious problem (although it *is* a problem), as the homomorphisms we have to deal with will usually be of a rather special type (centralizing, strongly normalizing or normalizing, for example – see Chapter II for definitions) and these behave acceptably with respect to prime ideals.

These homomorphisms also permit us to control the behaviour of radicals. For example, if I is a two-sided ideal of R and if φ is a centralizing extension, say, then $S \varphi (I) = J$ is a two-sided ideal of S and the image of σ_I in S–mod is exactly σ_J. However, if σ_I is stable, then σ_J is not necessarily stable in S–mod. So, even if there is a morphism $Spec(S) \to Spec(R)$, induced by $\varphi : R \to S$, it does not have to be continuous with respect to the above topology T_s.

A reason this fails (as we hope to make clear below) is that we at least need I to induce a stable radical for both left and right R-modules, a fact which leads us to consider biradicals in the sense of [44]. However, restricting the topology $T_s(R)$ to open subsets $X(I)$ such that I induces a biradical still does not solve our problems, as the image of a biradical is again not a biradical, in general.

One solution is to work with so-called centralizing biradicals, which may be defined by their nice relation with respect to centralizing bimodules, i.e., bimodules M generated over R by the elements $m \in M$ with the property that $rm = mr$ for all $r \in R$, cf. [3]. These centralizing biradicals *do* behave functorially, hence restricting our topology to the $X(I)$ such that σ_I is a centralizing biradical finally yields a complete answer, i.e., a functorial "affine scheme". Moreover, confining our attention to centralizing or somewhat more general biradicals is not as restrictive as one might think. Indeed, we will show that for rings satisfying the strong second layer condition (and our main examples are all of this type) centralizing biradicals are just ordinary biradicals and for radicals of the form σ_I this just amounts to I satisfying the left and right Artin-Rees condition. For full details, we refer to the text, of course.

Let us now very briefly describe the contents of this monograph. In Chapter I, we present the necessary background on abstract localization theory, used throughout this text. In particular, we will include some details on the second layer condition that will play a significant role in the sequel. In Chapter II, we study in some detail different types of ring extensions, concentrating mainly on features related to the behaviour of prime ideals and localization. Chapter III is completely devoted to the Artin-Rees property and its variants in the noncommutative case. Our main results show how these different versions are related, especially in the presence of the strong second layer condition. In the last chapter (which we intended to be rather open-ended, as an incentive to the creative reader), we study structure sheaves, after a thorough discussion of the various compatibility results needed to study their construction and functorial behaviour.

At this point, it is our pleasure to address some words of thanks to the persons who directly or indirectly contributed to this text: at the risk of forgetting many of them, let us mention Maribel Segura and Daniel Tarazona (for being there from the very beginning), Marie-Paule Malliavin (for showing a permanent interest in our work), Ken Brown and Tom Lenegan (responsible for big chunks of this text by the remark "You should take a look at the second layer condition," which they made at some point during a meeting in Antwerp), Arnold Beckelheimer (for many critical comments) and Freddy Van Oystaeyen (for continuous support and compatibility). We would also like to thank Pieter Verhaeghe, our local TeXpert, who helped us with layout and "artwork" and Eva Santos resp. Hugo Van Hove, for not being too angry when we used her resp. his office, while (noisily) discussing one of the many penultimate versions of this text in a mixture of English, French, Spanish and "Peperanto" [98]. Finally, we cannot overstress the influence of this other "ménage à trois", Luis Merino, Pepe Mulet, and Conchi Vidal, from whose Ph.D. theses we borrowed several results used in this text. We should also apologize to our beloved children, José, Maria del Mar, Natalia[1], Andrea, Laura, Maria[2], Noémie and Thomas[3], for spending too much time on this book.

We would like to dedicate this text to the memory of our good friend and colleague Pere Menal, who suddenly passed away, leaving us with a large, empty place in our hearts.

A final word of warning: this text is aimed at a mixed audience consisting of algebraists wanting to get some insight into sheaf-theoretic methods in noncommutative ring theory as well as algebraic geometers interested in noncommutative analogues of the objects they are used to working with. For this reason, this text contains material which should

---

[1]Supported by J. L. Bueso
[2]Supported by P. Jara
[3]Supported by A. Verschoren

be obvious to some readers, but probably not to others (e.g., the sections on primary decomposition and FBN rings resp. the introductory section on sheaves). The reader familiar with such topics should take a diagonal glance at the corresponding sections and just use them as a quick reminder or to fix notations.

Talking about notations: since we have to work frequently with both left and right modules, we always try to be as explicit as possible about the side we are using. For example, if M is a left R-module and I a two-sided ideal of R, then $Ann^r_M(I)$ will denote the set of all elements in M annihilated by I, i.e., the set of all $m \in M$ with $Im = 0$. Similarly, the left annihilator of I in R, i.e., the set of all $r \in R$ with $rI = 0$ will be denoted by $Ann^l_R(I)$. We should also stress the fact that inclusions will be denoted by $\subseteq$, reserving the symbol $\subset$ for *proper* inclusions. We write $I \leq R$ resp. $I \trianglelefteq R$ when I is a left or right resp. a two-sided ideal of R. For submodules, we will just use $\subseteq$ when no ambiguity arises. Finally, a noetherian ring will be left *and* right noetherian; the same terminology will be used for links, the second layer condition, and similar examples.

J. L. Bueso

P. Jara

A. Verschoren

Contents

PREFACE   v

I. LOCALIZATION   1

1. Localization at multiplicative subsets   2
2. Radicals 12
3. Localization 26
4. Symmetric radicals 39
5. Associated primes 47
6. The second layer condition 58
7. FBN rings 73

II. EXTENSIONS   81

1. Centralizing extensions 82
2. Normalizing extensions 90
3. Strongly normalizing extensions 98
4. Inducing radicals 105
5. Inducing localization functors 113
6. Functorial behaviour 120
7. Extensions and localization at prime ideals 128

III. STABILITY   143

1. Stable torsion theories 144

2. The Artin-Rees property 153

3. The weak Artin-Rees property 166

4. Biradicals 175

IV. COMPATIBILITY AND SHEAVES   187

1. Sheaves 189

2. The commutative theory 197

3. Compatibility in module categories 214

4. Topologies 232

5. Structure sheaves 238

BIBLIOGRAPHY   253

INDEX   263

Chapter I

LOCALIZATION

Is there anything more natural for a young child than asking about the quotient of 3 by 4, say? Clearly not, as one usually presents division as an inverse operation to multiplication. So, if 2 times 3 yields 6, conversely 6 by 2 returns 3. In other words, the first divisions a child is confronted with always terminate. Hence, why shouldn't there be a quotient of 3 by 4? Well, of course, there *is* a quotient of 3 by 4, but the point is: it isn't an integer anymore!

So, let us introduce fractions. Although this answers the child's need[1] for a quotient of 3 by 4, one should not forget that the seemingly innocent action of embedding the integers into the rationals really destroys a lot. Actually, allowing arbitrary quotients hides almost everything that makes integers so attractive to work with. OK, you obtain a nice, cozy field to play with, but what you lose is the fact that the integers themselves already form a euclidean ring, a unique factorization domain, a Dedekind ring with trivial class group, etc.

Let us not exaggerate. Everyone reading this text is, of course, well aware of the benefits of the introduction of fields of fractions and the fact of being able, through it, to apply Galois theory, algebraic geome-

---

[1]Children really want to know about fractions – at least, so do Andrea, José, Laura, Maria, Maria del Mar, Natalia, Noémie and Thomas.

try, ... to solve number-theoretic problems. So it is really not necessary for us to stress the necessity of introducing localization techniques in commutative and noncommutative algebra and ring theory.

We have tried to make this first Chapter into a toolkit, which mainly aims at making the techniques and results in the subsequent Chapters more accessible to the reader. For this reason, one should not expect here a comprehensive introduction to abstract localization theory in the sense of [27, et al] (albeit in a slightly more intuitive language), as we did not strive at completeness or self-containedness in any way. Quite the opposite: what we present of localization theory will just consist of the results (some of them unpublished or undocumented in the literature), which are needed in order to start reading the other Chapters. For full details (and missing proofs), we refer to the bibliography at the end.

1. Localization at multiplicative subsets

(1.1) So, how does one construct fractions? Let us stick to the commutative case for a moment. When working with integers, to introduce fractions, one essentially just formally inverts nonzero elements, i.e., one just adds an inverse $c^{-1} = 1/c$ to any integer c and one lets $r/c = c^{-1}r$ stand for the product of c^{-1} by r. Of course, one should be aware of the fact that r/c and s/d represent the same element, when $rd = sc$.

For an arbitrary commutative domain R, one proceeds in exactly the same way. One starts from a subset C of R, which is multiplicatively closed. In other words, C should contain $1 \in R$ (in order to be able to embed R into $C^{-1}R$) and if $c, d \in C$, then cd should also belong to C (in order to obtain a *ring* of fractions). One then formally introduces elements of the form $c^{-1}r$, with $r \in R$ and $c \in C$, and one again identifies $c^{-1}r$ and $d^{-1}s$ if $cs = dr$. Multiplication and addition are defined exactly as for the integers. This yields a domain, denoted by $C^{-1}R$, and mapping $r \in R$ to $r/1 = 1^{-1}r \in C^{-1}R$ embeds R into the ring of fractions $C^{-1}R$.

If R contains zero-divisors, one has to be a little more careful, i.e., one should identify $c^{-1}r$ and $d^{-1}s$, exactly when $e(cs - dr) = 0$ for some $e \in C$.

(1.2) Somewhat disappointingly, it appears that the analogous construction does not always work in the noncommutative case. In fact, one is interested in finding, for an arbitrary multiplicatively closed subset C of R a left *ring of fractions* of R with respect to C, i.e., a ring $C^{-1}R$ endowed with a ring homomorphism $j : R \to C^{-1}R$, (the so-called *localizing homomorphism*) with the properties that:

(1.2.1) $j(c)$ is invertible in $C^{-1}R$ for any $c \in C$;

(1.2.2) every element of $C^{-1}R$ is of the form $j(c)^{-1}j(r)$ for some $r \in R$ and some $c \in C$;

(1.2.3) we have $j(c)^{-1}j(r) = 0$ if and only if there exists some $d \in C$ with the property that $dr = 0$.

One may verify, cf. [52, 80], for example, that such a ring of fractions with respect to a multiplicatively closed subset C exists if and only if the following conditions hold:

1. (*(first) left Ore condition*) for any $r \in R$ and any $c \in C$, there exist $s \in R$ and $d \in C$ with the property that $dr = sc$;

2. (*left reversibility* or *second left Ore condition*) for any $r \in R$ and any $c \in C$ with the property that $rc = 0$, there exists some $d \in C$ such that $dr = 0$.

A multiplicatively closed subset C satisfying the previous properties will be called a *set of left denominators*; if it only satisfies the first left Ore condition, we say that it is a *left Ore set*.

Of course, in the noetherian case every left Ore set is automatically a set of left denominators:

(1.3) Lemma. *If the ring R is left noetherian, then the left Ore condition on a multiplicatively closed subset C in R implies left reversibility.*

Proof. Indeed, if $rc = 0$ for some $r \in R$ and $c \in C$, then the left noetherian assumption on R implies that $Ann_R^\ell(c^n) = Ann_R^\ell(c^{n+1})$ for some positive integer n. The left Ore condition implies that $sc^n = dr$ for some $s \in R$ and $d \in C$. It thus follows that $sc^{n+1} = drc = 0$. So, $s \in Ann_R^\ell(c^{n+1}) = Ann_R^\ell(c^n)$, i.e., $dr = sc^n = 0$. $\square$

Let us also point out the following useful left common denominator property, whose proof may be given by a straightforward induction argument:

(1.4) Lemma. *If C is a multiplicatively closed subset of R satisfying the left Ore condition and if $c_1, \ldots, c_n \in C$, then there exist $r_1, \ldots, r_n \in R$, such that $r_1c_1 = \ldots = r_nc_n \in C$.*

(1.5) The main result about localization at Ore sets is Goldie's Theorem [34], which in one of its many forms, says that for a prime noetherian ring R the set $\mathcal{C}_R$ of regular elements of R is a left Ore set. This permits to construct the simple artinian so-called *classical ring of fractions* $Q_{cl}(R) = \mathcal{C}_R^{-1}R$ of R.

The following Lemma, also due to Goldie, provides the main ingredient of its proof:

(1.6) Lemma. *[34, 37, 52, 80] If R is a left noetherian semiprime ring and L is a left ideal of R, then the following assertions are equivalent:*

(1.6.1) *L is essential in R (i.e., if I is a left ideal of R and $I \cap L = 0$, then $I = 0$);*

(1.6.2) *L contains a regular element.*

In fact, the previous Lemma immediately implies the left Ore condition on $\mathcal{C}_R$, for, if c is a regular element and if r is an arbitrary element in

R, then Rc and hence $(Rc : r)$ is easily seen to be essential. So, $(Rc : r)$ contains a regular element d, i.e., $dr = sc$ for some $s \in R$, indeed.

Left rings of fractions satisfy the following universal property:

(1.7) Lemma. *Let R be a ring, let C be a set of left denominators in R and denote by $j : R \to C^{-1}R$ the localizing homomorphism. Then, for any ring homomorphism $\varphi : R \to S$ with the property that $\varphi(c) \in S$ is invertible for any $c \in C$, there exists a unique ring homomorphism $\varphi' : C^{-1}R \to S$, with the property that $\varphi = \varphi' j$:*

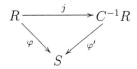

As a corollary, it follows that rings of fractions are unique (up to isomorphism) and that, moreover, if the left and right rings of fractions $C^{-1}R$ and RC^{-1} both exist, then these are isomorphic.

Let us already mention here the following useful result, whose (elementary) proof we leave to the reader (see also [37, Theorem 9.20.], for example):

(1.8) Lemma. *Let R be a left noetherian ring, C a left Ore set in R and I a twosided ideal of R. Then:*

(1.8.1) *the ring $C^{-1}R$ is left noetherian;*

(1.8.2) *$C^{-1}I = (C^{-1}R)I$ is a twosided ideal of $C^{-1}R$, and if $\overline{R} = R/I$ and*

$$\overline{C} = \{c + I; \ c \in C\},$$

then $\overline{C}$ is a left Ore set in $\overline{R}$ and $\overline{C}^{-1}\overline{R}$ is isomorphic to $C^{-1}R/C^{-1}I$;

(1.8.3) *if P is a prime ideal of R and if $P \cap C = \varnothing$, then $C^{-1}P$ is a prime ideal of $C^{-1}R$.*

(1.9) Corollary. [37, Theorem 9.22.] *Let R be a left noetherian ring and C a left Ore set. Then the map $P \mapsto C^{-1}P$ is an order preserving*

bijection between the prime ideals of R disjoint from C and the prime
ideals of $C^{-1}R$.

(1.10) The classical ring of fractions may be viewed as the noncommutative analogue of the field of fractions of a domain, i.e., as the localization of R at the prime ideal $\{0\}$. An obvious question should be:
how does one localize at an arbitrary prime ideal P of R? However, as
is well-known, although this works extremely well for the special prime
ideal $\{0\}$ of R (if R is prime and noetherian, of course), things do not
seem to work nicely at all for nonzero prime ideals.

Actually, although this remains outside of the scope of this text, even
the choice of which notion of a prime ideal to use is not trivial at all.
In fact, whereas $Spec(R)$, the set of all (twosided) prime ideals of R
appears to us to be the obvious one, (at least when we consider rings
possessing sufficiently many twosided ideals), one may also work efficiently with (isomorphism classes of) indecomposable injectives, prime
torsion theories, left prime ideals [71], irreducible ideals, completely
prime ideals, ..., cf. [32, 36, 47, 71, et al]

With respect to the last mentioned class, let us point out that the complement $R \setminus P$ of a completely prime ideal P of R is always multiplicatively closed, whereas this is not the case for an arbitrary (twosided)
prime ideal P. A possible remedy, if one wants to "localize at P" anyway, is to use a multiplicative subset of R, disjoint from P and which
coincides with $R \setminus P$, when P is completely prime.

The obvious candidate in this case appears to be the set $\mathcal{C}_R(P)$, which
consists of all elements of R, which are regular modulo P. For example,
if R is prime noetherian, then, with $P = 0$, we recover the Ore set $\mathcal{C}_R$
of regular elements of R as the set $\mathcal{C}_R(0)$.

(1.11) The next step, after the choice of this (hopefully) adequate
multiplicatively closed subset $\mathcal{C}_R(P)$, is to verify if a ring of fractions
with respect to it exists, i.e., to investigate whether $\mathcal{C}_R(P)$ is a set of

left denominators. It is exactly here that problems start to arise. In fact, not only is $\mathcal{C}_R(P)$ sometimes (or even usually!) not a set of left denominators, but there are many examples of rings, where $\mathcal{C}_R(P)$ is a set of left denominators for some prime ideals P of R, the so-called *localizable* ones, whereas it is not for others.

Of course, one is very interested in localizable prime ideals, due to their nice behaviour, which mimics (up to a certain degree) that of prime ideals in the commutative case. Indeed, if $P \in Spec(R)$ is a localizable prime ideal of the left noetherian ring R and if we denote by R_P the ring of fractions at the Ore set $\mathcal{C}_R(P)$, then (1.9) implies that there is a bijective correspondence between the prime ideals of R contained in P and the prime ideals of R_P. In particular, $R_P P$ is the unique maximal ideal of R_P. Moreover, Goldie's Theorem and (1.8) imply the quotient $R_P/R_P P$ to be isomorphic to the classical ring of fractions of R/P and it is left artinian as such.

(1.12) *For simplicity's sake, throughout the remainder of this Section, R will be assumed to be (left and right) noetherian.*

The bad behaviour of localization at a prime ideal is essentially due to two obstructions against localizability, which have been extensively studied in the literature:

(1.12.1) the existence of non-trivial links between prime ideals;

(1.12.2) the absence of the second layer property.

Let us briefly describe here some necessary background concerning these topics. We will not include any proofs or details at this point. Instead, the interested reader is referred to Section 6 for a more detailed treatment.

Consider prime ideals P, Q of R. If I is a twosided ideal of R, then we will say that there exists an *ideal link* from Q to P defined by I and we denote this by $Q \rightsquigarrow P$ (defined by I), if

(1.12.3) $QP \subseteq I \subset Q \cap P$;

(1.12.4) $Ann_R^\ell(Q \cap P/I) = Q$;

(1.12.5) $Ann_R^r(Q \cap P/I) = P$;

(1.12.6) $Q \cap P/I$ is torsionfree, both as a right R/P-module and as a left R/Q-module.

If $I = QP$, then we just say that there is a *link* from Q to P.

A subset $X \subseteq Spec(A)$ is said to be *right link closed* if for any $P \in X$ and any link $Q \rightsquigarrow P$ with $Q \in Spec(R)$, we have $Q \in X$. One defines in a similar way subsets $X \subseteq Spec(R)$ to be *left link closed* and *link closed* (on both sides). If $P \in Spec(R)$, then the *right clique* of P is denoted by $cl^r(P)$, and defined to be the smallest subset of $Spec(R)$ which contains P and which is right link closed. The *left clique* $cl^\ell(P)$ and the *clique* $cl(P)$ of P are defined similarly.

(1.13) Example. As a typical example, let us consider the (noetherian!) ring

$$\begin{pmatrix} \mathbf{Z} & \mathbf{Z} \\ 0 & \mathbf{Z} \end{pmatrix},$$

where $\mathbf{Z}$ is the ring of integers. It is fairly obvious that the prime ideals of R are exactly the ideals

$$P_p' = \begin{pmatrix} p\mathbf{Z} & \mathbf{Z} \\ 0 & \mathbf{Z} \end{pmatrix} \text{ resp. } P_p'' = \begin{pmatrix} \mathbf{Z} & \mathbf{Z} \\ 0 & p\mathbf{Z} \end{pmatrix},$$

where p is zero or a prime integer.

Since

$$P_p'P_p'' = \begin{pmatrix} p\mathbf{Z} & p\mathbf{Z} \\ 0 & p\mathbf{Z} \end{pmatrix} \subset \begin{pmatrix} p\mathbf{Z} & \mathbf{Z} \\ 0 & p\mathbf{Z} \end{pmatrix} = P_p' \cap P_p'' = P_p''P_p'$$

and since

$$P_p' \cap P_p''/P_p'P_p'' \cong \mathbf{Z}/p\mathbf{Z} \cong R/P_p' \cong R/P_p'',$$

it is easy to verify that for each p there is a link $P_p' \rightsquigarrow P_p''$ and *no* link $P_p'' \rightsquigarrow P_p'$.

It is also easy to see that for $p \neq 0$, there are links $P_p' \rightsquigarrow P_p'$ resp.

$P''_p \rightsquigarrow P''_p$ and that these are actually the only ones. All cliques over R thus consist of exactly two elements.

The following result shows how the presence of non-trivial links yields an obstruction against localizability at prime ideals, and explains why it makes more sense to localize with respect to cliques instead of just at single prime ideals.

(1.14) Lemma. [9, 37, 52, et al] *Let P be a prime ideal of R and let C be a left Ore set of R disjoint from P. Then*

$$C \subseteq \bigcap \{\mathcal{C}_R(Q); \ Q \in cl^{\ell}(P)\}.$$

In particular, the rather obvious idea of "localizing" at a prime ideal P by just considering the largest left Ore set contained within $\mathcal{C}_R(P)$ does not work, as this clearly does not permit to distinguish between different prime ideals in the left clique of P.

(1.15) As mentioned above, the second obstruction against localizability at a prime ideal is given by the absence of the second layer condition, which will be treated in full detail in Section 6 (in particular, see (6.8) and (6.9)).

Let us stress at this point that the class of rings satisfying the (rather technical) left second layer condition is surprisingly vast. Indeed, in the following noetherian rings (whose list is not exhaustive), every prime ideal satisfies the second layer condition, cf. [9, 10, 11, 37, 45, 52, 74, 75, et al]:

(1.15.1) fully bounded noetherian (FBN) rings, e.g., noetherian pi rings;

(1.15.2) artinian rings;

(1.15.3) principal ideal rings;

(1.15.4) hereditary noetherian prime (HNP) rings with enough invertible ideals;

(1.15.5) group rings RG and enveloping algebras $R \otimes U(\mathfrak{g})$, where R is a noncommutative noetherian ring, G a polycyclic-by-finite group, $\mathfrak{g}$ is a finite dimensional Lie (super) algebra, which is assumed to be solvable in characteristic 0;

(1.15.6) Ore extensions of the form $R[\theta; \delta]$, $R[x, \varphi]$ and $R[x, x^{-1}, \phi]$, where R is a commutative noetherian ring, δ a derivation on R and φ an automorphism of R;

(1.15.7) the group-graded and the skew-enveloping analogues of the previous types of rings.

Note that if $\mathfrak{g}$ is a *non-solvable* Lie algebra over a field of characteristic 0 and $R = U(\mathfrak{g})$ or if R is a hereditary noetherian prime ring *without* enough invertible ideals, then R has prime ideals, which do not satisfy the left or right second layer condition.

Let us also point out that, with the exception of the general group-graded and skew enveloping cases, all of the above examples satisfy the so-called *strong* second layer condition.

Finally, let us mention that it has been proved, by E.S. Letzter [49], that for any pair of noetherian rings $R \subseteq S$ such that S is left and right finitely generated as an R-module, the ring S satisfies the (strong) second layer condition, whenever R does.

(1.16) Let $X \subseteq Spec(R)$, then we say that X is *left localizable*, if there exists a left Ore set $C \subseteq R$ with $P \cap C = \varnothing$ for all $P \in X$, and such that

(1.16.1) for all $P \in X$, the ring $C^{-1}R/C^{-1}P$ is artinian;

(1.16.2) the primitive ideals of $C^{-1}R$ are exactly the ideals of the form $C^{-1}P$, with $P \in X$.

We will say that X is *classically left localizable* if it also satisfies the property that every finitely generated left $C^{-1}R$-module which is an essential extension of a simple $C^{-1}R$-module is artinian.

Let us already point out here, that if X is left localizable, then X is classically left localizable if and only if X satisfies the left second layer condition.

For any $X \subseteq Spec(R)$, we let

$$\mathcal{C}_R(X) = \bigcap\{\mathcal{C}_R(P);\ P \in X\}.$$

It is then clear that if X is left localizable, then $C \subseteq \mathcal{C}_R(X)$ and $\mathcal{C}_R(X)$ is a left Ore set. Moreover, there exists a ring isomorphism $C^{-1}R \cong \mathcal{C}_R(X)^{-1}R$. One usually denotes this ring by R_X and calls it the (left) *localization* or *ring of fractions* of R at X.

The following Theorem is one of the main results about classical localizability (for unexplained terminology and a proof, we refer to [9] or (6.9)):

(1.17) Theorem. [9, 37, 52, et al] *Let R be a (noetherian) ring and $X \subseteq Spec(R)$ a set of prime ideals of R. Then X is classically left localizable if and only if:*

(1.17.1) *X is left link closed;*

(1.17.2) *X satisfies the left second layer condition;*

(1.17.3) *X satisfies the weak left intersection property;*

(1.17.4) *X satisfies the incomparability property.*

It now finally follows, by taking $X = \{P\}$:

(1.18) Corollary. *A prime ideal P of a (noetherian) ring R is classically left localizable if and only if it satisfies the left second layer condition and if $cl^\ell(P) = \{P\}$.*

2. Radicals

(2.1) Let R be an arbitrary ring and let $R-\mathbf{mod}$ denote the category of left R-modules. Consider an additive subfunctor

$$\sigma : R-\mathbf{mod} \to R-\mathbf{mod}$$

of the identity in $R-\mathbf{mod}$, i.e., for all $M \in R-\mathbf{mod}$, one has $\sigma M \subseteq M$ and $f(\sigma M) \subseteq \sigma M'$ for any homomorphism $f : M \to M'$ of left R-modules.

We will call[2] the functor σ *idempotent* resp. a *preradical*, if $\sigma(\sigma M) = \sigma M$ resp. $\sigma(M/\sigma M) = 0$, for any $M \in R-\mathbf{mod}$. A left exact preradical is said to be a *radical* and is sometimes also referred to as an *idempotent kernel functor*, in the sense of Goldman [35].

(2.2) Proposition. *Any left exact subfunctor σ of the identity in $R-\mathbf{mod}$ is idempotent.*

Proof. Let M be a left R-module. As $\sigma M \subseteq M$, it clearly follows from the left exactness of σ that

$$\sigma(\sigma M) = Ker(\sigma M \to \sigma(M/\sigma M))$$
$$= Ker(\sigma M \to M/\sigma M) = \sigma M,$$

which proves that σ is idempotent, indeed.   □

(2.3) Proposition. *Let σ be a preradical in $R-\mathbf{mod}$. If M is a left R-module and L a submodule of σM, then $\sigma(M/L) = \sigma M/L$.*

---

[2]The reader should be warned about the fact that the terminology we adopt in this text differs from the one in [80], for example. Actually, what we call a preradical resp. radical is frequently referred to as a radical resp. left exact preradical. We prefer the present terminology, due to the fact that starting from the next section, all radicals (in the terminology of [80]) will be left exact.

Proof. The quotient morphism $M \to M/L$ induces a morphism $\sigma M \to \sigma(M/L)$ with kernel L. It follows that $\sigma M/L \subseteq \sigma(M/L)$. Conversely, since $\sigma(M/\sigma M) = 0$, the canonical map $p : M/L \to M/\sigma M$ vanishes on $\sigma(M/L)$, hence $\sigma(M/L) \subseteq Ker(p) = \sigma M/L$. $\quad\square$

(2.4) Examples.

(2.4.1) Let R be an integral domain. For any R-module M, let us denote by

$$t(M) = \{m \in M;\ \exists\, 0 \neq r \in R, rm = 0\}$$

the torsion of M. The functor

$$t : R-\mathbf{mod} \to R-\mathbf{mod} : M \mapsto t(M)$$

is a radical.

(2.4.2) Somewhat more generally, let R be an arbitrary commutative ring and let C be a multiplicatively closed subset of R, or, even more generally, let R be arbitrary, but assume C to be a set of left denominators in R. For any $M \in R-\mathbf{mod}$, put

$$\sigma_C M = \{m \in M;\ \exists c \in C, cm = 0\}.$$

It is again easy to see that the functor

$$\sigma_C : R-\mathbf{mod} \to R-\mathbf{mod} : M \mapsto \sigma_C M$$

is a radical.

In particular, if R is commutative and if $\mathfrak{p} \in Spec(R)$, then $C = R \setminus \mathfrak{p}$ is multiplicatively closed, so this defines a radical $\sigma_{R\setminus\mathfrak{p}}$ in $R-\mathbf{mod}$.

(2.4.3) Let R be an arbitrary ring and let I be a twosided ideal of R. Let M be a left R-module, then we put

$$\sigma_I M = \{m \in M;\ \exists n \in \mathbf{N}, I^n m = 0\}.$$

This defines a left exact idempotent functor

$$\sigma_I : R-\mathbf{mod} \to R-\mathbf{mod} : M \mapsto \sigma_I M,$$

in $R-\mathbf{mod}$. If R is left noetherian, or, more generally, if we just assume I to be finitely generated as a left R-ideal, then σ_I is a radical.

(2.4.4) Let R be an arbitrary ring. For any left R-module M, let us define the socle $S(M)$ of M by

$$S(M) = \sum\{N \subseteq M; \ N \ is \ simple\}$$

The functor

$$S : R-\mathbf{mod} \to R-\mathbf{mod} : M \mapsto S(M)$$

is left exact, but not necessarily a preradical.

(2.4.5) Let R be an arbitrary ring. For any left R-module M, let us define by

$$J(M) = \bigcap\{N \subseteq M; \ N \ maximal \ in \ M\}$$

the Jacobson radical of M. The functor

$$J : R-\mathbf{mod} \to R-\mathbf{mod} : M \mapsto J(M)$$

is a preradical, which, in general, is neither idempotent nor left exact.

(2.4.6) Let R be an arbitrary ring. For any left R-module M, define by

$$Z(M) = \{x \in M; \ Ann_R^\ell(x) \ is \ essential \ in \ R\}$$

the singular submodule of M. The functor

$$Z : R-\mathbf{mod} \to R-\mathbf{mod} : M \mapsto Z(M)$$

is left exact (hence idempotent), but not a preradical, in general.

(2.4.7) Let R be a commutative domain. An R-module M is said to be *divisible* if $rM = M$ for any $0 \neq r \in R$. Put $\sigma M = \sum M'$, where M' runs through all divisible R-submodules of M, i.e., σM is the largest divisible R-submodule of M. It is easy to see that this defines an idempotent radical σ, which is not left exact in general, however.

(2.5) A pair of non-empty classes of left R-modules $(\mathcal{T}, \mathcal{F})$ is said to be a *torsion theory* for $R-\mathbf{mod}$ if

(2.5.1) $\mathcal{T} = \{M \in R-\mathbf{mod};\ Hom_R(M, N) = 0\ for\ all\ N \in \mathcal{F}\}$;

(2.5.2) $\mathcal{F} = \{N \in R-\mathbf{mod};\ Hom_R(M, N) = 0\ for\ all\ M \in \mathcal{T}\}$.

In particular, it then follows that $\mathcal{T} \cap \mathcal{F} = \{0\}$.

If $(\mathcal{T}, \mathcal{F})$ is a torsion theory for $R-\mathbf{mod}$, then the modules in $\mathcal{T}$ are said to be *torsion*, whereas those in $\mathcal{F}$ are said to be *torsionfree*.

(2.6) Proposition. *Let R be an arbitrary ring, then:*

(2.6.1) *a non-empty class $\mathcal{T}$ of left R-modules is the class of torsion modules for a torsion theory $(\mathcal{T}, \mathcal{F})$ if and only if $\mathcal{T}$ is closed under taking extensions, direct sums and quotients;*

(2.6.2) *a non-empty class $\mathcal{F}$ of left R-modules is the class of torsionfree modules for a torsion theory $(\mathcal{T}, \mathcal{F})$ if and only if $\mathcal{F}$ is closed under taking extensions, direct products and submodules.*

Proof. Although this result belongs to torsion-theoretic folklore, (cf. [80], for example), let us include the proof here, for the reader's convenience.

If $M \in \mathcal{T}$ and if M maps onto N, say by a surjection $M \to N$, then for any $F \in \mathcal{F}$, we have $Hom_R(N, F) \hookrightarrow Hom_R(M, F) = 0$, so $N \in \mathcal{T}$, as well. It is also clear that if $\{M_i;\ i \in I\}$ is a family of torsion modules, then

$$Hom_R(\bigoplus_{i \in I} M_i, F) = \prod_{i \in I} Hom_R(M_i, F) = 0,$$

so $\bigoplus_i M_i \in \mathcal{T}$. Finally, if $M \in R-\mathbf{mod}$ fits into an exact sequence

$$0 \to M' \to M \to M'' \to 0,$$

with $M', M'' \in \mathcal{T}$, then any morphism $f \in Hom_R(M, F)$ with $F \in \mathcal{F}$, vanishes on M', hence factorizes over M'', say as

$$
\begin{array}{ccc}
M & \xrightarrow{\ p\ } & M'' \\
{\scriptstyle f}\downarrow & \swarrow {\scriptstyle \bar{f}} & \\
F & &
\end{array}
$$

Since $M'' \in \mathcal{T}$, we find that $\bar{f} = 0$, however, so $f = 0$. Hence $Hom_R(M, F) = 0$ and $M \in \mathcal{F}$, indeed.

Conversely, assume that $\varnothing \neq \mathcal{T} \subseteq R-\mathbf{mod}$ is closed under taking extensions, direct sums and quotients. Let $(\mathcal{T}', \mathcal{F})$ denote the torsion theory generated by $\mathcal{T}$, i.e., put

$$
\mathcal{F} = \{N \in R-\mathbf{mod};\ \forall M \in \mathcal{T}, Hom_R(M, N) = 0\}
$$

and

$$
\mathcal{T}' = \{M \in R-\mathbf{mod};\ \forall N \in \mathcal{F}, Hom_R(M, N) = 0\},
$$

(it is trivial to see that $(\mathcal{T}', \mathcal{F})$ is a torsion theory, indeed).

To finish the proof, let us verify that $\mathcal{T} = \mathcal{T}'$. Obviously, $\mathcal{T} \subseteq \mathcal{T}'$. Conversely, let $M \in \mathcal{T}'$, i.e., assume that $Hom_R(M, F) = 0$ for all $F \in \mathcal{F}$, then we want to prove that $M \in \mathcal{T}$. As $\mathcal{T}$ is closed under taking direct sums and quotients, there obviously exists a largest left R-submodule $N \subseteq M$ belonging to $\mathcal{T}$ (the sum of all left R-submodules of M that belong to $\mathcal{T}$!). Let us show that $M/N \in \mathcal{F}$. This will imply that $M = N$ and finish the proof. Assume not and pick a nonzero morphism $f : M'' \to M/N$ with $M'' \in \mathcal{T}$. The image of f also belongs to $\mathcal{T}$ and its inverse image in M through $M \to M/N$ is a left R-submodule of M which belongs to $\mathcal{T}$ (as $\mathcal{T}$ is closed under extensions) and which strictly contains N. As this contradicts the maximality of N, this proves the assertion.

The proof of the second assertion may be given in a similar way. $\square$

(2.7) Proposition. *Let $(\mathcal{T}, \mathcal{F})$ be a pair of non-empty classes of left R-modules, then $(\mathcal{T}, \mathcal{F})$ is a torsion theory if and only if the following* *conditions hold:*

(**2.7.1**) $\mathcal{T} \cap \mathcal{F} = \{0\}$;

(**2.7.2**) *for every* $M \in R-\textbf{mod}$ *there exists a submodule* σM *such that* $\sigma M \in \mathcal{T}$ *and* $M/\sigma M \in \mathcal{F}$.

Proof. That any torsion theory $(\mathcal{T}, \mathcal{F})$ satisfies the conditions of the statement (with σM the sum of all torsion submodules of M, i.e., the largest torsion left R-submodule of M) has essentially been proved in the previous result.

Conversely, assume that $(\mathcal{T}, \mathcal{F})$ satisfies the above conditions. Suppose that $M \in R-\textbf{mod}$ and that $Hom_R(M, N) = 0$ for all $N \in \mathcal{F}$. Then, in particular, the canonical surjection $M \to M/\sigma M$ vanishes, i.e., $M = \sigma M \in \mathcal{T}$.

Similarly, let $N \in R-\textbf{mod}$ and assume $Hom_R(M, N) = 0$ for all $M \in \mathcal{T}$, then, in particular, the canonical inclusion $\sigma N \hookrightarrow N$ vanishes, showing that $\sigma N = 0$ and that $N = N/\sigma N \in \mathcal{F}$. This proves the assertion. $\square$

(**2.8**) **Proposition.** *Let* $(\mathcal{T}, \mathcal{F})$ *be a torsion theory over* $R-\textbf{mod}$. *For any left* R-*module* M, *put* $\sigma M = \sum \{N \subseteq M; \ N \in \mathcal{T}\}$. *Then:*

(**2.8.1**) *if* $f \in Hom_R(M, N)$, *then* $f(\sigma M) \subseteq \sigma N$;

(**2.8.2**) $\sigma(M/\sigma M) = 0$;

(**2.8.3**) $\sigma(\sigma M) = \sigma M$.

In other words, σ *is an idempotent preradical.*

Proof. The first statement is obvious, the second follows essentially from the fact that $\mathcal{T}$ is closed under taking quotients and the last one from the fact that $\mathcal{T}$ is closed both under taking quotients and direct sums.

Finally, if $\sigma(M/\sigma M) \neq 0$, then, as $\mathcal{T}$ is closed under taking extensions, its inverse image in M would yield, as before, a torsion left R-submodule of M, strictly containing σM – a contradiction. $\square$

(2.9) Proposition. *Let σ be an idempotent preradical in $R-$mod and let us define*

$$\mathcal{T}_\sigma = \{M \in R-\text{mod}; \ \sigma M = M\}$$

resp.

$$\mathcal{F}_\sigma = \{M \in R-\text{mod}; \ \sigma M = 0\}.$$

Then $(\mathcal{T}_\sigma, \mathcal{F}_\sigma)$ is a torsion theory for $R-$mod.

Proof. Obviously, $\mathcal{T}_\sigma \cap \mathcal{F}_\sigma = \{0\}$. On the other hand, for any $M \in R-$mod, clearly $\sigma M \in \mathcal{T}_\sigma$ (as σ is idempotent) and $M/\sigma M \in \mathcal{F}_\sigma$ (as σ is a preradical). Applying (2.7) immediately yields the result. $\square$

(2.10) Proposition. *The following sets correspond bijectively:*

(2.10.1) *torsion theories over $R-$mod;*

(2.10.2) *idempotent preradicals in $R-$mod.*

Proof. This follows easily from the previous results. $\square$

(2.11) Proposition. *Let $(\mathcal{T}, \mathcal{F})$ be a torsion theory for $R-$mod. If σ is the idempotent preradical associated to $(\mathcal{T}, \mathcal{F})$, then the following assertions are equivalent:*

(2.11.1) $\mathcal{T}$ *is closed under taking submodules;*

(2.11.2) $\mathcal{F}$ *is closed under taking injective hulls.*

Proof. First assume that $\mathcal{T}$ is closed under taking submodules. Let $N \in \mathcal{F}$ and let $E(N)$ be its injective hull. Denote by σ the idempotent preradical associated to $(\mathcal{T}, \mathcal{F})$, then $\sigma E(N) \cap N \in \mathcal{T} = \mathcal{T}_\sigma$, as a submodule of $\sigma E(N)$, hence $\sigma E(N) \cap N \subseteq \sigma N = 0$. So, $\sigma E(N) = 0$, as $E(N)$ is essential over N. This proves that (1) implies (2).

Conversely, assume that $\mathcal{F}$ is closed under taking injective hulls. Let $M \in \mathcal{T}$ and let N be a left R-submodule of M, then there exists some morphism $q : M \to E(N/\sigma N)$, extending the morphism

$$N \xrightarrow{p} N/\sigma N \hookrightarrow E(N/\sigma N).$$

Since $E(N/\sigma N) \in \mathcal{F}$ and $M \in \mathcal{T}$, it follows that $q = 0$, hence $p = 0$, as well, i.e., $N = \sigma N \in \mathcal{T}$. This proves that (2) implies (1). $\qquad\square$

A torsion theory which satisfies the previous equivalent conditions is said to be *hereditary*.

(2.12) Theorem. *There exists a bijective correspondence between*

(2.12.1) *hereditary torsion theories in $R-\mathbf{mod}$;*

(2.12.2) *radicals in $R-\mathbf{mod}$.*

Proof. In view of the foregoing discussion, it only remains to be verified that an idempotent radical σ is left exact if and only if its torsion class $\mathcal{T}_\sigma$ is closed under taking submodules. One implication being obvious, assume that $\mathcal{T}_\sigma$ is closed under taking R-submodules and let $N \subseteq M$ be left R-modules. Obviously, $\sigma N \subseteq \sigma M \cap N$. Conversely, $\sigma M \cap N \in \mathcal{T}_\sigma$, as a left R-submodule of σM, so $\sigma M \cap N \subseteq \sigma N$ (whence equality), since σ is idempotent. This shows that σ is left exact, indeed, and finishes the proof. $\qquad\square$

(2.13) Example. Let C be an arbitrary multiplicatively closed subset of a ring R, then we may define a hereditary torsion theory in $R-\mathbf{mod}$ by letting its torsion class $\mathcal{T}_C$ consists of all left R-modules with the property that for all $m \in M$, there exists some $c \in C$ with $cm = 0$.

The associated radical in $R-\mathbf{mod}$ is denoted by σ_C. It has been noted in [45, (1.1.6)] that C is a left Ore set if and only if $R/Rc \in \mathcal{T}_C$ for all $c \in C$ or, equivalently, if

$$\sigma_C M = \{m \in M;\ \exists c \in C, cm = 0\}$$

for all $M \in R-\mathbf{mod}$, cf. (2.4.2).

As a special case, for any prime ideal P of R, this defines a radical $\sigma_P = \sigma_{\mathcal{C}(P)}$ in $R-\mathbf{mod}$, where $\mathcal{C}(P)$ is the multiplicatively closed subset associated to P in the same way as before.

Note that it has been proved in [47] that if R is left noetherian, then σ_P may also be defined on any left R-module M by

$$\sigma_P = \bigcap \{Ker(f); \ f \in Hom_R(M, E(R/P))\},$$

where $E(R/P)$ denotes an injective hull of R/P in $R-\mathbf{mod}$.

(2.14) Although the torsionfree class $\mathcal{F}$ of a hereditary torsion theory $(\mathcal{T}, \mathcal{F})$ in $R-\mathbf{mod}$ is closed under taking injective hulls, this is not necessarily true for its torsion class $\mathcal{T}$. If it is, then we will say that $(\mathcal{T}, \mathcal{F})$ and the associated left exact radical σ in $R-\mathbf{mod}$ are *stable*. Of course, this is also equivalent to asserting that any essential extension of a σ-torsion left R-module is σ-torsion.

The concept of stability appears to be a very natural (and useful) one. In particular, let us already point out that over a noetherian commutative ring, for example, *every* left exact radical is stable, as we will see below.

(2.15) Let us now briefly describe another way of efficiently dealing with the concept of torsion. Recall that a non-empty set $\mathcal{L}$ of left ideals of R is said to be a *Gabriel filter* [27] if it satisfies the following conditions:

(2.15.1) if $I \in \mathcal{L}$ and $a \in R$, then $(I : a) \in \mathcal{L}$;

(2.15.2) if I is a left ideal of R for which there exists some $J \in \mathcal{L}$ such that $(I : a) \in \mathcal{L}$ for any $a \in J$, then $I \in \mathcal{L}$.

The reason why we speak of a Gabriel *filter* is motivated by:

(2.16) Proposition. *If $\mathcal{L}$ is a Gabriel filter of left ideals of R, then the following conditions are satisfied:*

(2.16.1) *if $I \in \mathcal{L}$ and if J is a left ideal of R with the property that $I \subseteq J$, then $J \in \mathcal{L}$;*

(2.16.2) *if $I, J \in \mathcal{L}$, then $I \cap J \in \mathcal{L}$.*

Proof. First note that $R \in \mathcal{L}$, as for any $I \in \mathcal{L}$ (which is non-empty), we have $R = (I : 0) \in \mathcal{L}$ by (2.15.1). So, if $I \subseteq J$, we find for any $a \in I$ that $(J : a) = R \in \mathcal{L}$, hence, by (2.15.2), that $J \in \mathcal{L}$, indeed. On the other hand, if I and J both belong to $\mathcal{L}$ and if $a \in I$, then

$$(I \cap J : a) = (J : a) \in \mathcal{L},$$

by (2.15.1), hence $I \cap J \in \mathcal{L}$, by (2.15.2). $\square$

(2.17) Example. Let R be an arbitrary ring and let I be a twosided ideal of R, which is finitely generated as a left ideal. Then

$$\mathcal{L}(I) = \{L \leq R; \ \exists n \in \mathbf{N}, I^n \subseteq L\}$$

is a Gabriel filter.

Indeed, if $L \in \mathcal{L}(I)$, then $L \supseteq I^n$ for some positive integer n. So, if $a \in R$, then from $I^n a \subseteq I^n \subseteq L$, it follows that $I^n \subseteq (L : a)$, so $(L : a) \in \mathcal{L}(I)$, indeed.

Next, let L be a left ideal of R and assume that there exists some $J \in \mathcal{L}(I)$, which we may pick of the form $J = I^n$ for some positive integer n, with $(L : a) \in \mathcal{L}(I)$ for all $a \in J = I^n$. Then, for each $a \in I^n$, there exists some positive integer $m(a)$ with $I^{m(a)} \subseteq (L : a)$, i.e., $I^{m(a)} a \subseteq L$. As I^n is easily verified to be finitely generated as a left ideal, there exists some positive integer m with $I^m a \subseteq L$ for all $a \in I^n$, i.e., $I^{m+n} \subseteq L$. So, L belongs to $\mathcal{L}(I)$, indeed.

A subset $\mathcal{B}$ of $\mathcal{L}$ is said to be a *basis* for $\mathcal{L}$ if for any $I \in \mathcal{L}$, we may find some $J \in \mathcal{B}$ with $J \subseteq I$. If $\mathcal{L}$ has a basis consisting of finitely generated left ideals, then we say that $\mathcal{L}$ has *finite type*. In particular, the filter $\mathcal{L}(I)$ just defined is an example of a Gabriel filter of finite type.

(2.18) Corollary. *Let $\mathcal{L}$ be a Gabriel filter of left ideals of R. If both I and J belong to $\mathcal{L}$, then so does IJ.*

Proof. Clearly $I \subseteq (IJ : J)$, hence the result follows from (2.15.2) and (2.16.1). $\square$

Note also that for commutative rings, the definition of a Gabriel filter may be somewhat simplified:

(2.19) Lemma. *Let R be a commutative ring and let $\mathcal{L}$ be a family of ideals of R. Then $\mathcal{L}$ is a Gabriel filter, if and only if*

(2.19.1) $R \in \mathcal{L}$;

(2.19.2) *if I is an ideal of R for which there exists some $J \in \mathcal{L}$ such that $(I : a) \in \mathcal{L}$ for all $a \in J$, then $I \in \mathcal{L}$.*

Proof. Note that the second condition is just (2.15.2) in the definition of a Gabriel filter. So, let us first assume $\mathcal{L}$ to be a Gabriel filter. Then $\mathcal{L}$ is non-empty, so picking $I \in \mathcal{L}$ and $a = 0$ in (2.15.1) shows that $R = (I : a) \in \mathcal{L}$. This proves that any Gabriel filter satisfies (1) and (2). Conversely, assume that (1) and (2) hold true, then we just have to prove (2.15.1) in order to verify that $\mathcal{L}$ is a Gabriel filter. First note that if $J \in \mathcal{L}$ and if $J \subseteq I$, then $I \in \mathcal{L}$. Indeed, this follows from (2) and the fact that for any $a \in J$ we have $(I : a) = R \in \mathcal{L}$. To derive (2.15.1), it now suffices to note that for any $I \in \mathcal{L}$ and $a \in R$, we have $I \subseteq (I : a)$.   □

(2.20) Proposition. *Let σ be a radical over $R-\mathrm{mod}$. Then*

$$\mathcal{L}(\sigma) = \{I \leq R; \ R/I \in \mathcal{T}_\sigma\}$$

is a Gabriel filter over R.

Proof. It is clear that $\mathcal{L}(\sigma)$ is non-empty, since $R \in \mathcal{L}(\sigma)$. Next, let $I \in \mathcal{L}(\sigma)$ and pick $a \in R$. Consider the multiplication morphism

$$\mu_a : R \to R : r \mapsto ra,$$

then, clearly, $\mu_a^{-1}(I) = (I : a)$. So, μ_a induces an injective morphism

$$\overline{\mu_a} : R/(I : a) \hookrightarrow R/I.$$

Since R/I is σ-torsion, by assumption, so is $R/(I : a)$, i.e., $(I : a) \in \mathcal{L}(\sigma)$.

Finally, let I be a left R-ideal and assume that for some $J \in \mathcal{L}(\sigma)$ we have $(I : a) \in \mathcal{L}(\sigma)$, for all $a \in J$. Then we want to show that $I \in \mathcal{L}(\sigma)$, as well. Consider the following exact sequence:

$$0 \to J/I \cap J \to R/I \to R/I + J \to 0.$$

Since $R/I + J$ is a quotient of R/J, it is σ-torsion. On the other hand, $J/I \cap J$ is σ-torsion as well, since $(I \cap J : j) = (I : j) \in \mathcal{L}(\sigma)$, for any $j \in J$. It thus follows that R/I is σ-torsion, hence that $I \in \mathcal{L}(\sigma)$, indeed.   □

(2.21) Proposition. *Let $\mathcal{L}$ be a Gabriel filter for R. Then*

$$\sigma_{\mathcal{L}} M = \{x \in M;\ \exists I \in \mathcal{L}, Ix = 0\}$$

determines a radical $\sigma_{\mathcal{L}}$ in $R-\mathbf{mod}$.

Proof. It is fairly easy to see that $\sigma_{\mathcal{L}}$ is a left exact subfunctor of the identity in $R-\mathbf{mod}$.

On the other hand, let M be a left R-module and consider $\overline{m} \in \sigma_{\mathcal{L}}(M/\sigma_{\mathcal{L}} M)$ for some $m \in M$. By definition, there exists some $J \in \mathcal{L}$ such that $Jm \subseteq \sigma_{\mathcal{L}} M$. Let $a \in J$, then $am \in \sigma_{\mathcal{L}} M$, so there exists some $I_a \in \mathcal{L}$ with the property that $I_a(am) = 0$. Hence $I_a \subseteq (Ann_R^{\ell}(m) : a)$ and it follows that $(Ann_R^{\ell}(m) : a) \in \mathcal{L}$. As this holds for all $a \in J$, we thus obtain that $Ann_R^{\ell}(m) \in \mathcal{L}$, indeed.   □

(2.22) Proposition. *The following sets correspond bijectively:*

(2.22.1) *Gabriel filters for R;*

(2.22.2) *radicals in $R-\mathbf{mod}$.*

Proof. This is an easy consequence of the previous results.   □

If the Gabriel filter $\mathcal{L}$ has finite type, then we will also say that the associated radical $\sigma_{\mathcal{L}}$ has *finite type*.

As a straightforward example, for any left finitely generated twosided ideal I of R, it is easy to see that the Gabriel filter $\mathcal{L}(I)$ and the radical σ_I correspond to each other under the above bijections and that σ_I thus has finite type.

Note also:

(2.23) Lemma. *Let σ be a radical on $R-$mod. The following assertions are equivalent:*

(2.23.1) *σ has finite type;*

(2.23.2) *σ commutes with inductive limits.*

Proof. To show that (1) implies (2), let us consider an inductive system of left R-modules $\{M_\alpha;\ \alpha \in A\}$. Since the torsion class $\mathcal{T}_\sigma$ is closed under taking direct sums and quotients, we get an obvious inclusion $\varinjlim \sigma M_\alpha \subseteq \sigma(\varinjlim M_\alpha)$. On the other hand, if we take $m \in \sigma(\varinjlim M_\alpha)$, then $Lm = 0$ for some finitely generated left ideal $L \in \mathcal{L}(\sigma)$. But then, we may represent m by some $m_\alpha \in M_\alpha$ for a suitably chosen α, such that we still have $Lm_\alpha = 0$, i.e., $m_\alpha \in \sigma M_\alpha$. So, $m \in \varinjlim \sigma M_\alpha$.

Conversely, if $L \in \mathcal{L}(\sigma)$, then we may write L as a directed union of finitely generated left ideals L_α. Since $R/L = \varinjlim R/L_\alpha$ is σ-torsion, we find $R/L = \sigma(R/L) = \varinjlim \sigma(R/L_\alpha)$. In particular, since $\bar{1} \in R/L$ belongs to this inductive limit, we may pick some $r \in R$ and some $K \in \mathcal{L}(\sigma)$, such that $1 - r \in L$ and $Kr \subseteq L_\alpha$. Choosing α with the property that $1 - r \in L_\alpha$, it follows that $K \subseteq L_\alpha$, hence $L_\alpha \in \mathcal{L}(\sigma)$, indeed. $\square$

(2.24) Let us finish this Section with some remarks on the lattice structure of the set $R-$**rad** of radicals in $R-$**mod**.

First, a partial order $\leq$ on $R-$**rad** is defined, by letting $\sigma \leq \tau$ if and only if $\sigma M \subseteq \tau M$ for any $M \in R-$**mod**. For example, if R is a commutative ring, I a finitely generated ideal of R and $I \not\subseteq \mathfrak{p}$

a prime ideal of R, then (with notations as before), we clearly have $\sigma_I \leq \sigma_{\mathfrak{p}} = \sigma_{R \backslash \mathfrak{p}}$. It is easy to see that this is equivalent to $\mathcal{T}_\sigma \subseteq \mathcal{T}_\tau$ and to $\mathcal{F}_\tau \subseteq \mathcal{F}_\sigma$. Moreover, if $G \subseteq R-\mathbf{rad}$ is a family of radicals, then the *meet* $\bigwedge G$ and the *join* $\bigvee G$ of G are given by

$$\mathcal{T}_{\bigwedge G} = \bigcap_{\sigma \in G} \mathcal{T}_\sigma$$

resp.

$$\mathcal{F}_{\bigvee G} = \bigcap_{\sigma \in G} \mathcal{F}_\sigma.$$

With these definitions, it is then fairly easy to verify that $R-\mathbf{rad}$ is a complete distributive lattice.

Note also that for any $M \in R-\mathbf{mod}$ we have

$$(\bigwedge G)M = \bigcap_{\sigma \in G} \sigma M \text{ resp. } (\bigvee G)M \supseteq \sum_{\sigma \in G} \sigma M,$$

the second inclusion possibly being strict. In particular, we then also have

$$\mathcal{L}(\bigwedge G) = \bigcap_{\sigma \in G} \mathcal{L}(\sigma),$$

whereas an analogous description for $\mathcal{L}(\bigvee G)$ does not exist, in general – see (4.5), however.

3. Localization

(3.1) Let σ be a radical in $R-\mathbf{mod}$ and let E be a left R-module. We say that E is σ-injective if it has the property that for any pair of left R-modules $M' \subseteq M$ with $M/M' \in \mathcal{T}_\sigma$, the canonical map

$$Hom_R(M, E) \to Hom_R(M', E)$$

is surjective. In other words, any left R-linear map $f : M' \to E$ should extend to a left R-linear map $\overline{f} : M \to E$:

$$
\begin{array}{ccc}
M' & \longrightarrow & M \\
{\scriptstyle f}\downarrow & \swarrow{\scriptstyle \overline{f}} & \\
E & &
\end{array}
$$

If $\overline{f}$ is unique as such, then we say that E is *faithfully σ-injective* or *σ-closed*.

Essentially the same proof as that of Baer's Lemma shows:

(3.2) Proposition. *A left R-module E is (faithfully) σ-injective if and only if for any $L \in \mathcal{L}(\sigma)$ and any left R-linear map $f : L \to E$, there exists some (unique) $e \in E$ with the property that $f(l) = le$ for any $l \in L$.*

We thus want any $f : L \to E$ to extend (uniquely) to $\overline{f} : R \to E$:

$$
\begin{array}{ccc}
L & \longrightarrow & R \\
{\scriptstyle f}\downarrow & \swarrow{\scriptstyle \overline{f}} & \\
E & &
\end{array}
$$

We also have the following characterization of σ-closed R-modules:

(3.3) Proposition. *Let E be a left R-module. Then the following assertions are equivalent:*

(3.3.1) *E is σ-closed;*

(3.3.2) *E is σ-injective and σ-torsionfree.*

Proof. If E is σ-closed, then E is certainly σ-injective. On the other hand, by definition, the inclusion $\sigma E \hookrightarrow E$ is the unique extension of the zero map $0 \hookrightarrow E$, so $\sigma E = 0$, indeed.

Conversely, assume that E is σ-injective and σ-torsionfree. Consider $N \subseteq M$ with $M/N \in \mathcal{T}_\sigma$. Any left R-linear map $f : N \to E$ then extends to some left R-linear map $f : M \to E$. Assume there is another extension $f'' : M \to E$, then $g = f' - f''$ vanishes on N, hence factorizes through $\overline{g} : M/N \to E$. However, since $M/N \in \mathcal{T}_\sigma$ and $E \in \mathcal{F}_\sigma$, clearly $\overline{g} = 0$, hence $f' = f''$. This proves that the extension f' is unique, i.e., that E is σ-closed, indeed. $\qquad\square$

Let us call a left R-linear map $u : N \to M$ a *σ-isomorphism* if both $Ker(u)$ and $Coker(u)$ are σ-torsion.

Then:

(3.4) Proposition. *If E is a σ-closed left R-module, then for any σ-isomorphism $u : N \to M$, the canonical map*

$$Hom_R(u, E) : Hom_R(M, E) \to Hom_R(N, E)$$

is bijective.

Proof. Since $Ker(u) \in \mathcal{T}_\sigma$, clearly any $f \in Hom_R(N, E)$ vanishes on $Ker(u)$, hence uniquely factorizes over $N/Ker(u)$. So, the map

$$Hom_R(N, E) \to Hom_R(N/Ker(u), E)$$

is bijective.
On the other hand, since

$$Coker(N/Ker(u) \to M) = Coker(u) \in \mathcal{T}_\sigma,$$

the map

$$Hom_R(N/Ker(u), E) \to Hom_R(M, E)$$

is bijective as well, by assumption. Composing the previous bijections yields the assertion. $\qquad\square$

(**3.5**) Let us denote by $(R, \sigma)-\mathbf{mod}$ the full subcategory of $R-\mathbf{mod}$, consisting of σ-closed left R-modules. It is fairly easy to see that it is canonically endowed with a Grothendieck category structure. Let us point out that the kernel of a map between σ-closed left R-modules is σ-closed, whereas this is not true, in general, for cokernels. In fact, a morphism $f : M \to N$ in $(R, \sigma)-\mathbf{mod}$ is an epimorphism if and only if $Coker(f)$ (its cokernel in $R-\mathbf{mod}$) is σ-torsion.

We wish to define an exact functor $Q_\sigma : R-\mathbf{mod} \to (R, \sigma)-\mathbf{mod}$, which is left adjoint to the inclusion $i_\sigma : (R, \sigma)-\mathbf{mod} \hookrightarrow R-\mathbf{mod}$ (a so-called *reflector* of $R-\mathbf{mod}$ in $(R, \sigma)-\mathbf{mod}$).

(**3.6**) For any $M \in R-\mathbf{mod}$, let us denote by $E_\sigma(M)$ the left R-submodule of $E(M)$, an injective hull of M, consisting of all $m \in E(M)$ with $Lm \subseteq M$ for some $L \in \mathcal{L}(\sigma)$. So, $E_\sigma(M) = p^{-1}(\sigma(E(M)/M))$, where $p : E(M) \to E(M)/M$ is the canonical surjection. In particular, $E(M)/E_\sigma(M) \in \mathcal{F}_\sigma$ and $E_\sigma(M)/M \in \mathcal{T}_\sigma$.

Let us define the *localization* or *left R-module of quotients* $Q_\sigma(M)$ of a left R-module M at σ by $Q_\sigma(M) = E_\sigma(M/\sigma M)$. The localization $Q_\sigma(M)$ is canonically endowed with a left R-linear map

$$j_{\sigma,M} : M \to M/\sigma M \hookrightarrow E_\sigma(M/\sigma M) = Q_\sigma(M).$$

The next result implies that $Q_\sigma(M) \in (R, \sigma)-\mathbf{mod}$ for any $M \in R-\mathbf{mod}$:

(**3.7**) **Lemma.** *If $M \in \mathcal{F}_\sigma$, then $E_\sigma(M)$ is σ-closed.*

Proof. First note that $E_\sigma(M)$ is certainly σ-torsionfree, being a left R-submodule of $E(M) \in \mathcal{F}_\sigma$. Next, let $I \in \mathcal{L}(\sigma)$ and consider a left R-linear map $f : I \to E_\sigma(M)$. Clearly, f extends to a left R-linear map $\overline{f} : R \to E(M)$, which fits into a commutative square

$$
\begin{array}{ccc}
I & \xrightarrow{\ \ i\ \ } & R \\
{\scriptstyle f}\downarrow & & \downarrow{\scriptstyle \overline{f}} \\
E_\sigma(M) & \xrightarrow[\ \ j\ \ ]{} & E(M)
\end{array}
$$

Let $p : E(M) \to E(M)/E_\sigma(M)$ denote the quotient morphism, then $p\overline{f}i = pjf = 0$, so $\overline{f}$ factorizes through $f' : R/I \to E(M)/E_\sigma(M)$. But $R/I \in \mathcal{T}_\sigma$ and $E(M)/E_\sigma(M) \in \mathcal{F}_\sigma$, so f' is the zero morphism. Hence $\overline{f}$ actually maps R into $E_\sigma(M) \subseteq E(M)$, which proves that $E_\sigma(M)$ is σ-injective as well.   □

Note also that $E_\sigma(M) = M$, if M is σ-closed. Indeed, if M is σ-closed, then (since $E_\sigma(M)/M \in \mathcal{T}_\sigma$), the identity on M extends to a left R-linear map $t : E_\sigma(M) \to M$, which is automatically surjective. To see that t is also injective, consider its kernel $K = Ker(t)$ and assume that $K \neq 0$. As $K \subseteq E_\sigma(M) \subseteq E(M)$, obviously $K \cap M \neq 0$, which would imply the existence of some $0 \neq m \in M$ with $t(m) = 0$. As t extends the identity on M, this yields a contradiction.

We may now prove:

(3.8) Proposition. *The previous construction defines an idempotent left exact functor* $Q_\sigma : R-\mathbf{mod} \to R-\mathbf{mod}$.

Proof. Let $f : M \to N$ be a left R-linear map, then, since $j_{\sigma,M} : M \to Q_\sigma(M)$ is a σ-isomorphism and since $Q_\sigma(N)$ is σ-closed, f uniquely extends to $Q_\sigma(f) : Q_\sigma(M) \to Q_\sigma(N)$ fitting into the commutative diagram

$$
\begin{array}{ccc}
M & \xrightarrow{\ \ f\ \ } & N \\
{\scriptstyle j_{\sigma,M}}\downarrow & & \downarrow{\scriptstyle j_{\sigma,N}} \\
Q_\sigma(M) & \xrightarrow[Q_\sigma(f)]{} & Q_\sigma(N)
\end{array}
$$

We leave it to the reader to verify that the uniqueness of $Q_\sigma(f)$ implies that this yields the desired functor Q_σ, indeed. The remarks preceding our assertion show Q_σ to be idempotent.   □

(3.9) From the previous result, it fairly easily follows that Q_σ maps σ-isomorphisms in $R-\mathbf{mod}$ to isomorphisms, essentially because Q_σ (by its very definition) vanishes on σ-torsion modules.

In particular, it follows that if $\sigma \leq \tau$, then $Q_\tau Q_\sigma = Q_\tau$. Indeed, since for any $M \in R-\mathbf{mod}$ the map $j_{\sigma,M} : M \to Q_\sigma(M)$ is a σ-isomorphism, it is also a τ-isomorphism, hence

$$Q_\tau(j_{\sigma,M}) : Q_\tau(M) \to Q_\tau(Q_\sigma(M))$$

is an isomorphism.

(3.10) Corollary. *The functor* Q_σ *defines an exact functor*

$$\mathbf{a}_\sigma : R-\mathbf{mod} \to (R,\sigma)-\mathbf{mod},$$

which is left adjoint to the inclusion $i_\sigma : (R,\sigma)-\mathbf{mod} \hookrightarrow R-\mathbf{mod}.$

Proof. Since Q_σ is idempotent and every $Q_\sigma(M)$ is σ-closed, clearly the image of Q_σ is exactly the full subcategory $(R,\sigma)-\mathbf{mod}$ of $R-\mathbf{mod}$. Exactness of the induced functor follows easily from the fact that Q_σ is left exact and the previously given description of epimorphisms in $(R,\sigma)-\mathbf{mod}$ (for details, see [80, 86], for example).

To show that $\mathbf{a}_\sigma$ is left adjoint to the inclusion i_σ, we have to verify for all $E \in (R,\sigma)-\mathbf{mod}$ and $M \in R-\mathbf{mod}$, that any $f \in Hom_R(M,E)$ corresponds bijectively to some $\overline{f} \in Hom_R(Q_\sigma(M),E)$. But this follows directly from the fact that E is σ-closed and $j_{\sigma,M} : M \to Q_\sigma(M)$ a σ-isomorphism. $\square$

The following result provides an alternative characterization of the localization functor:

(3.11) Lemma. *If* $M \in \mathcal{F}_\sigma$, *then*

$$Q_\sigma(M) = \varinjlim_{I \in \mathcal{L}(\sigma)} Hom_R(I, M),$$

up to canonical isomorphism.

Proof. Let us write

$$Q_{(\sigma)}(M) = \varinjlim_{I \in \mathcal{L}(\sigma)} Hom_R(I, M),$$

then $Q_{(\sigma)}(M)$ is endowed with a left R-module structure as follows: let $m \in Q_{(\sigma)}(M)$ be represented by $\mu : I \to M$ and let $r \in R$, then we define rm as the element in $Q_{(\sigma)}(M)$ represented by the composition

$$(I : r) \xrightarrow{r} I \xrightarrow{\mu} M.$$

There is a canonical map

$$h_M : M \to Q_{(\sigma)}(M),$$

which maps $m \in M$ to the class of $\mu_m : R \to M : r \mapsto rm$ and which is easily seen to be left R-linear with respect to the above left R-module structure.

First note that $Ker(h_M) = 0$. Indeed, if $m \in M$ and $h_M(m) = 0$, then the left R-linear homomorphism $R \to M : r \mapsto rm$ vanishes on some $L \in \mathcal{L}(\sigma)$, i.e., $Lm = 0$. Hence $m \in \sigma M = 0$.

On the other hand, clearly $Coker(h_M) \in \mathcal{T}_\sigma$. Indeed, somewhat more generally, we claim that if $m \in Q_{(\sigma)}(M)$ is represented by $\mu : I \to M$ with $I \in \mathcal{L}(\sigma)$, then the diagram

$$
\begin{array}{ccc}
I & \longrightarrow & R \\
\mu \downarrow & & \downarrow \varphi_m \\
M & \xrightarrow{h_M} & Q_{(\sigma)}(M)
\end{array}
$$

(where $\varphi_m(r) = rm$ for any $r \in R$) commutes. This follows immediately from the fact that if $i \in I$, then im is represented by

$$R = (I : i) \xrightarrow{i} I \xrightarrow{\mu} M : r \mapsto \mu(ri) = r\mu(i).$$

We thus have proved that h_M is a σ-isomorphism. To conclude, it thus suffices to show that $Q_{(\sigma)}(M)$ is σ-closed. To verify that $Q_{(\sigma)}(M)$ is σ-torsionfree, assume that $Lm = 0$ for some $m \in Q_{(\sigma)}(M)$ and $L \in \mathcal{L}(\sigma)$. Let m be represented by $\mu : I \to M$ with $I \in \mathcal{L}(\sigma)$. Then there is a commutative diagram

$$
\begin{array}{ccc}
I & \longrightarrow & R \\
{\scriptstyle\mu}\downarrow & & \downarrow{\scriptstyle\varphi_M} \\
M & \xrightarrow[h_M]{} & Q_{(\sigma)}(M)
\end{array}
$$

So, $h_M\mu|L \cap I = 0$, hence $\mu|L \cap I = 0$, as h_M is injective. Since $L \cap I \in \mathcal{L}(\sigma)$, clearly $\mu|L \cap I$ also represents m, i.e., $m = 0$. It thus follows that $Q_{(\sigma)}(M) \in \mathcal{F}_\sigma$, indeed.

Next, let us show that $Q_{(\sigma)}(M)$ is σ-injective. Start from $f : L \to Q_{(\sigma)}(M)$ with $L \in \mathcal{L}(\sigma)$ and let $I = f^{-1}(M)$, then we obtain an exact commutative diagram

$$
\begin{array}{ccccccccc}
0 & \longrightarrow & I & \longrightarrow & L & \longrightarrow & L/I & \longrightarrow & 0 \\
& & {\scriptstyle f|I}\downarrow & & {\scriptstyle f}\downarrow & & \downarrow{\scriptstyle g} & & \\
0 & \longrightarrow & M & \xrightarrow[h_M]{} & Q_{(\sigma)}(M) & \longrightarrow & Coker(h_M) & \longrightarrow & 0
\end{array}
$$

Since g is monomorphic and $Coker(h_M) \in \mathcal{T}_\sigma$, we obtain $L/I \in \mathcal{T}_\sigma$, hence $I \in \mathcal{L}(\sigma)$. Now, as pointed out before, $f|I$ fits into a commutative square

$$
\begin{array}{ccc}
I & \longrightarrow & R \\
{\scriptstyle f|I}\downarrow & & \downarrow{\scriptstyle\varphi_m} \\
M & \xrightarrow[h_M]{} & Q_{(\sigma)}(M)
\end{array}
$$

where m is the class of $f|I$ in $Q_{(\sigma)}(M)$.

But then φ_m also extends f (which proves the assertion). Indeed, $\varphi_m|I = f|I$ implies that $(\varphi_m|L - f)|I = 0$, so $\varphi_m|L - f$ factorizes through $L/I \in \mathcal{T}_\sigma$. As $Q_{(\sigma)}(M) \in \mathcal{F}_\sigma$, this shows this factorization to be the zero map, hence $\varphi_m|L = f$, indeed.   □

(3.12) Note. It is fairly easy to see directly that

$$
\varinjlim_{I \in \mathcal{L}(\sigma)} Hom_R(I, M) = 0,
$$

if $M \in \mathcal{T}_\sigma$. Indeed, choose any left R-linear homomorphism $f : I \to M$ with $I \in \mathcal{L}(\sigma)$, representing an element μ in this inductive limit. Then $J = Ker(f) \in \mathcal{L}(\sigma)$, as I/J injects into M, which is σ-torsion. But then the restriction of f to J, i.e., the zero map $J \to M$, still represents μ, hence $\mu = 0$, indeed.

(3.13) Corollary. *For any left R-module M, we have*

$$Q_\sigma(M) = \varinjlim_{I \in \mathcal{L}(\sigma)} Hom_R(I, M/\sigma M),$$

up to canonical isomorphism.

(3.14) Proposition. *(cf. [80], for example) Consider a radical σ in $R-\mathbf{mod}$. Then:*

(3.14.1) *the R-module $Q_\sigma(R)$ is canonically endowed with a ring structure, extending that of R;*

(3.14.2) *any σ-closed left R-module is a left $Q_\sigma(R)$-module, in a canonical way.*

If M happens to be an R-bimodule, more may be said, cf. [87]:

(3.15) Proposition. *Let M be an R-bimodule and let σ be a radical in $R-\mathbf{mod}$. Then $Q_\sigma(M)$ is canonically endowed with a (unique) R-bimodule structure, extending that of M.*

Proof. Since σM is an R-subbimodule of M, upon replacing M by $M/\sigma M$, we may clearly assume M to be σ-torsionfree.

Let $j_{\sigma,M} : M \to Q_\sigma(M)$ denote the canonical inclusion. Pick $r \in R$ and denote by φ_r the left R-linear morphism

$$\varphi_r : M \xrightarrow{\cdot r} M \xrightarrow{j_{\sigma,M}} Q_\sigma(M) : m \mapsto mr.$$

Since $Q_\sigma(M)$ is σ-closed, clearly φ_r uniquely extends to some left R-linear morphism $\overline{\varphi_r} : Q_\sigma(M) \to Q_\sigma(M)$. We now define the right

R-action on $Q_\sigma(M)$ by putting $mr = \overline{\varphi_r}(m)$ for any $m \in Q_\sigma(M)$ and any $r \in R$. Obviously,

$$(sm)r = \overline{\varphi_r}(sm) = s\overline{\varphi_r}(m) = s(mr).$$

As $\overline{\varphi_{r+s}}$ and $\overline{\varphi_r} + \overline{\varphi_s}$ coincide, when restricted to M, a straightforward unicity argument yields that $m(r + s) = mr + ms$, for any $m \in Q_\sigma(M)$ and $r, s \in R$. Similarly, $\overline{\varphi_{rs}} = \overline{\varphi_r}\,\overline{\varphi_s}$, hence $m(rs) = (mr)s$, for all $m \in Q_\sigma(M)$ and $r, s \in R$.

So, $Q_\sigma(M)$ is endowed with an R-bimodule structure, extending that of M.

Conversely, if $Q_\sigma(M)$ is endowed with an R-bimodule structure which extends that of M, then right multiplication by $r \in R$ in $Q_\sigma(M)$ is left R-linear and induces φ_r on M. Another unicity argument then yields that the R-bimodule structure defined on $Q_\sigma(M)$ in the above way coincides with the given one, hence is actually unique. $\square$

In fact, it is fairly easy to see that $Q_\sigma(M)$, endowed with this R-bimodule structure may be viewed as the localization of M at σ in the Grothendieck category of R-bimodules, cf. [87].

Note also that, although $Q_\sigma(M)$ is a left $Q_\sigma(R)$-module, it is only a left R-module, and not necessarily a right $Q_\sigma(R)$-module, in general.

(3.16) Although localization at radicals appears to behave roughly as localization with respect to multiplicative sets (in the commutative case) or, more generally, at sets (in the noncommutative case), there are some big differences. Indeed, even in the commutative case, localization at a radical σ is (in general) not exact, does not commute with direct sums and the localization map $j_{\sigma,R} : R \to Q_\sigma(R)$ is neither left flat nor epimorphic, just to mention a few examples.

It appears, somewhat surprisingly, that these properties (or better: the lack of them) are tightly connected. Indeed, we have the following result, due to Goldman [35, Theorem 4.3.]:

(3.17) Theorem. [35] *Let σ be a radical, then the following assertions are equivalent:*

(3.17.1) *every left $Q_\sigma(R)$-module is σ-torsionfree;*

(3.17.2) *for every $L \in \mathcal{L}(\sigma)$ we have $Q_\sigma(R)j_{\sigma,R}(L) = Q_\sigma(R)$;*

(3.17.3) *every left $Q_\sigma(R)$-module is faithfully σ-injective;*

(3.17.4) *for any left R-module M, the canonical map*

$$Q_\sigma(R) \otimes_R M \to Q_\sigma(M)$$

is an isomorphism;

(3.17.5) *the functor Q_σ is exact and commutes with direct sums.*

Proof. Let us include the proof, for the reader's convenience. In order to simplify notations, we will assume throughout R to be σ-torsionfree. This allows us to identify the localization map $j_{\sigma,R}$ with the canonical inclusion $R \subseteq Q_\sigma(R)$.

To prove that (1) implies (2), note that in the exact sequence

$$0 \to R/L \to Q_\sigma(R)/L \to Q_\sigma(R)/R \to 0$$

both extremes are torsion, Hence so is the middle term $Q_\sigma(R)/L$. But then $Q_\sigma(R)/Q_\sigma(R)L$ is σ-torsion as well. Since our assumption also implies it to be σ-torsionfree, it follows that $Q_\sigma(R)L = Q_\sigma(R)$, indeed. Let us now assume (2) and let M be a left $Q_\sigma(R)$-module. First note that $M \in \mathcal{F}_\sigma$. Indeed, if $m \in M$ and if $L \in \mathcal{L}(\sigma)$ with $Im = 0$, then $Q_\sigma(R)m = Q_\sigma(R)Lm = 0$, so $m = 0$. Next, to prove that M is σ-injective, consider a left R-linear map $f : L \to M$, which we want to extend to R. From (2), it follows that we may write $1 = \sum_i q_i l_i$, for some $q_i \in Q_\sigma(R)$ and $l_i \in L$. Pick $K \in \mathcal{L}(\sigma)$ with $Kq_i \subseteq R$ for all i, then $H = K \cap L \in \mathcal{L}(\sigma)$ as well. Let $x = \sum_i q_i f(l_i) \in M$, then

$$hx = \sum_i hq_i f(l_i) = \sum_i f(hq_i l_i) = f(h),$$

for any $h \in H$, (as $hq_i \in R$). This shows that $f|H$ extends to $\overline{f} : R \to M$. Since $L/H \in \mathcal{T}_\sigma$, it easily follows from this that $\overline{f}$ also extends f, thus proving that M is σ-injective. This shows that (2) implies (3).

Let us now prove that (3) implies (4). First note that $Q_\sigma(R) \otimes_R M$ is, in the obvious way, a left $Q_\sigma(R)$-module, for any $M \in R-\mathbf{mod}$, so it is σ-closed, by assumption. As $Q_\sigma(R) \otimes_R M$ is, in particular, σ-torsionfree, it follows that the kernel $Ker(u)$ of the canonical homomorphism $u : M \to Q_\sigma(R) \otimes_R M$ contains σM. Moreover, if $\mu : Q_\sigma(R) \otimes_R M \to Q_\sigma(M)$ denotes the multiplication map, it follows that the composition μu is just the localization map $j_{\sigma,M} : M \to Q_\sigma(M)$, so we actually have $Ker(u) = \sigma M$.

On the other hand, we have

$$Coker(u) = Q_\sigma(R) \otimes_R M/u(M) \in \mathcal{T}_\sigma,$$

so u is a σ-isomorphism. Localizing at σ thus induces an isomorphism

$$Q_\sigma(M) \cong Q_\sigma(Q_\sigma(R) \otimes_R M) = Q_\sigma(R) \otimes_R M,$$

which proves the assertion.

It is obvious that (4) implies (5). As we have already seen in the proof of the implication (2) $\Rightarrow$ (3) that (2) implies (1), we now show that (5) yields (2), to finish the proof.

If $L \in \mathcal{L}(\sigma)$, then $Q_\sigma(R)L$ is a left $Q_\sigma(R)$-module. Choose a surjective map $u : A \to Q_\sigma(R)L$, where A is a free left $Q_\sigma(R)$-module, say of the form $A = Q_\sigma(R)^{(I)}$, for some index set I. Since Q_σ is assumed to commute with direct sums, clearly $A = Q_\sigma(R^{(I)})$, hence, in particular, A is then σ-closed. Since Q_σ is also assumed to be right exact, u induces a surjective map $A = Q_\sigma(A) \to Q_\sigma(Q_\sigma(R)L)$. Since $Q_\sigma(R)L$ is σ-torsionfree, it follows that $Q_\sigma(Q_\sigma(R)L) = Q_\sigma(R)L$, so $Q_\sigma(R)L$ is σ-closed. But, since we have shown in the proof of (1) $\Rightarrow$ (2) that $Q_\sigma(R)/Q_\sigma(R)L \in \mathcal{T}_\sigma$, we thus obtain that $Q_\sigma(R)L = Q_\sigma(R)$, indeed.

$\square$

A radical σ satisfying one (and hence all) of the conditions in the previous result is said to be *perfect* [27, 80] or to satisfy *property (T)* [35]. It follows from the previous result that when σ is perfect, then the localization at σ may be viewed as an exact functor

$$Q_\sigma : R-\mathrm{mod} \to Q_\sigma(R)-\mathrm{mod} = (R, \sigma)-\mathrm{mod},$$

which behaves in most ways like traditional localization at multiplicatively closed subsets of commutative rings. For more details, we refer to [31, 33, 35, 80, 83, et al].

(3.18) Example. If R is a Dedekind domain, then *every* radical σ in $R-\mathrm{mod}$ has property (T). Indeed, we claim that for any $L \in \mathcal{L}(\sigma)$ we have $Q_\sigma(R)I = Q_\sigma(R)$. Actually, since R is a Dedekind domain, L is invertible, say with inverse

$$L^{-1} = \{q \in K; \ Lq \subseteq R\},$$

where K is the field of fractions of R.

Since $LL^{-1} \subseteq R$ and since $L \in \mathcal{L}(\sigma)$, clearly $L^{-1} \subseteq Q_\sigma(R)$. But then $Q_\sigma(R)L \supseteq L^{-1}L = R$, proving that $1 \in Q_\sigma(R)L$, hence that $Q_\sigma(R)L = Q_\sigma(R)$, as claimed.

(3.19) Note. If the radical σ has property (T), then σ automatically has finite type. Indeed, assume R to be σ-torsionfree, for notational convenience and let $L \in \mathcal{L}(\sigma)$. Then $Q_\sigma(R)L = Q_\sigma(R)$, so $1 = \sum_{i=1}^n q_i l_i$ for some $q_i \in Q_\sigma(R)$ and some $l_i \in L$. Let $K = \sum_{i=1}^n Rl_i \subseteq L$, then clearly $Q_\sigma(R)K = Q_\sigma(R)$. So,

$$Q_\sigma(R/K) = Q_\sigma(R) \otimes_R R/K = Q_\sigma(R)/Q_\sigma(R)K = 0,$$

hence $R/K \in \mathcal{T}_\sigma$ or, equivalently, $K \in \mathcal{L}(\sigma)$. As K is finitely generated, this proves our claim.

On the other hand, let us point out that Q_σ may well be exact, without σ being of finite type (hence without σ having property (T)). Indeed,

consider the following example, essentially due to F. Call [22].
Let k be an arbitrary field and consider the ring $R \subseteq k^N$ of eventually
constant sequences of elements in k. Let e_i denote the sequence that
has 1 in the ith position and 0 elsewhere, and put $\mathfrak{m} = \bigoplus_i R e_i$. Then
$\mathfrak{m}$ is the maximal ideal of R consisting of all sequences that eventually
become zero and clearly $\mathfrak{m}^2 = \mathfrak{m}$. So, although $\mathfrak{m}$ is obviously *not*
finitely generated, we may associate to it a radical $\sigma_\mathfrak{m}$ in $R-mod$, with
torsion class $\mathcal{T}_\mathfrak{m}$ consisting of all R-modules M with the property that
$\mathfrak{m}M = 0$. It is easy to verify that $\mathcal{L}(\sigma_\mathfrak{m}) = \{\mathfrak{m}, R\}$. Hence $\sigma_\mathfrak{m}$ is
certainly not of finite type, as $\mathfrak{m}$ is not finitely generated.

On the other hand, we claim that the associated localization functor
$Q_\mathfrak{m}$ in $R-mod$ is exact. Indeed, since $\mathcal{L}(\sigma_\mathfrak{m})$ contains a minimal ideal
(i.e., $\mathfrak{m}$) and since R is clearly torsionfree at $\sigma_\mathfrak{m}$, we find that

$$Q_\mathfrak{m}(R) = Hom_R(\mathfrak{m}, R) = Hom_R(\bigoplus_i Re_i, R)$$
$$= \prod_i Hom_R(Re_i, R) = \prod_i Re_i \cong k^N.$$

Similarly, if M is an arbitrary R-module and if $\overline{M} = M/\sigma_\mathfrak{m}M$, then
$Q_\mathfrak{m}(M) = \prod_i e_i \overline{M}$. Finally, note that $e_i\overline{M} \cong e_iM$. Indeed, the canonical surjection $M \to \overline{M}$ induces a surjection $e_iM \to e_i\overline{M}$, that is actually
also injective. For, if $e_im = n \in e_iM$ with $\overline{n} = \overline{0}$ in $e_i\overline{M}$, then $n \in \sigma_\mathfrak{m}M$,
so $0 = e_in = e_i^2m = e_im = n$. We thus find that $Q_\mathfrak{m}(M) = \prod_i e_iM$ (up
to canonical isomorphism) and it follows from this description that the
localization functor $Q_\mathfrak{m}$ is exact as claimed.

4.   Symmetric radicals

(4.1)   A radical σ in R-**mod** is said to be *symmetric* [83] or *bounded* [27], if the set $\mathcal{L}^2(\sigma)$ of twosided ideals in $\mathcal{L}(\sigma)$ is a filter basis for $\mathcal{L}(\sigma)$. In other words, σ is symmetric if every left ideal in $\mathcal{L}(\sigma)$ contains a twosided ideal, which still belongs to $\mathcal{L}(\sigma)$.

Of course, if σ is symmetric, then the associated (hereditary) torsion theory $(\mathcal{T}, \mathcal{F})$ will also be called symmetric. We will denote the subset of $R-\textbf{rad}$ consisting of symmetric torsion theories by $R-\textbf{sym}$.

(4.2) Lemma.   *A radical σ is symmetric if and only if for any left ideal L of R the following assertions are equivalent:*

(4.2.1) $L \in \mathcal{L}(\sigma)$;

(4.2.2) $(L : R) \in \mathcal{L}(\sigma)$.

Proof.   Let us first assume σ to be symmetric. If L is a left ideal of R, then $(L : R)$ is the largest twosided ideal of R contained in L. If $(L : R) \in \mathcal{L}(\sigma)$, then clearly also $L \in \mathcal{L}(\sigma)$. Conversely, any $L \in \mathcal{L}(\sigma)$ contains a twosided ideal I of R, with $I \in \mathcal{L}(\sigma)$. So, $I \subseteq (L : R)$ hence $(L : R) \in \mathcal{L}(\sigma)$.

Conversely, the equivalence of (1) and (2) implies that any left ideal of in $\mathcal{L}(\sigma)$ contains a twosided ideal in $\mathcal{L}(\sigma)$. $\square$

Since for any symmetric radical σ in $R-\textbf{mod}$ the filter $\mathcal{L}^2(\sigma)$ is cofinal in $\mathcal{L}(\sigma)$, it immediately follows:

(4.3) Proposition.   *If σ is symmetric radical, then*

$$Q_\sigma(M) \cong \varinjlim_{I \in \mathcal{L}^2(\sigma)} Hom_R(I, M/\sigma M),$$

for any left R-module M.

(4.4)   If σ is a radical in $R-\textbf{mod}$, then, by (2.18), the filters $\mathcal{L}(\sigma)$ and $\mathcal{L}^2(\sigma)$ are multiplicatively closed. Conversely, if R is left noetherian, then any multiplicatively closed filter $\mathcal{L}$ of twosided R-ideals defines a (symmetric) radical $\sigma = \sigma_{\mathcal{L}}$ in $R-\mathbf{mod}$, by putting

$$\sigma_{\mathcal{L}} M = \{ m \in M; \; \exists I \in \mathcal{L}, Im = 0 \},$$

for any left R-module M.

Indeed, first note that $\mathcal{L}$ is closed under taking finite intersections, as for any pair of twosided R-ideals I, J we have $IJ \subseteq I \cap J$. From this it easily follows that σ actually yields a left exact subfunctor of the identity in $R-\mathbf{mod}$. It thus remains to verify that $\sigma(M/\sigma M) = 0$ for any $M \in R-\mathbf{mod}$.

Let $m \in M$ and denote by $\overline{m}$ its image in $M/\sigma M$. If $\overline{m} \in \sigma(M/\sigma M)$, then $I\overline{m} = \overline{0}$ for some $I \in \mathcal{L}$, i.e., $Im \subseteq \sigma M$. As R is left noetherian, I is finitely generated as a left R-ideal, say $I = \sum_{\alpha=1}^{n} Ri_{\alpha}$ for some $i_{\alpha} \in I$. For each $1 \leq \alpha \leq n$, we may pick some $J_{\alpha} \in \mathcal{L}$ with $J_{\alpha} i_{\alpha} m = 0$, so, with $J = \bigcap_{\alpha=1}^{n} J_{\alpha} \in \mathcal{L}$, we obtain $J i_{\alpha} m = 0$ for all α, hence $JIm = 0$. Since $\mathcal{L}$ is multiplicatively closed, we get $m \in \sigma M$, i.e., $\overline{m} = \overline{0}$.

As a consequence of the previous remarks, we obtain:

(4.5) Proposition. *Let R be left noetherian and let $G \subseteq R-\mathbf{sym}$ be a family of symmetric radicals in $R-\mathbf{mod}$. Then, for any left ideal L of R, the following assertions are equivalent:*

(4.5.1) *$L \in \mathcal{L}(\bigvee G)$;*

(4.5.2) *there exists a finite number of left R-ideals $L_1, \ldots, L_n$ in $\bigcup_{\sigma \in G} \mathcal{L}(\sigma)$ with the property that $L \supseteq \prod_{\alpha=1}^{n} I_{\alpha}$.*

Proof. Since the radicals in G are symmetric, we may obviously restrict to *twosided* ideals in the second statement. So, denote by $\mathcal{L}$ the filter of twosided ideals of R containing some finite product $\prod_{\alpha=1}^{n} I_{\alpha}$, where the I_{α} belong to $\bigcup_{\sigma \in G} \mathcal{L}^2(\sigma)$. As $\mathcal{L}$ is multiplicatively closed, it defines a symmetric radical τ in $R-\mathbf{mod}$, with the property that $\mathcal{L} = \mathcal{L}^2(\tau)$. It is also obvious that we have $\mathcal{L}(\sigma) \subseteq \mathcal{L}(\tau)$, i.e., $\sigma \leq \tau$, for any $\sigma \in G$, so $\bigvee G \leq \tau$.

Finally, let us assume λ is a radical in $R-\mathbf{mod}$ with the property that $\sigma \leq \lambda$ for all $\sigma \in G$, i.e., $\mathcal{L}(\sigma) \subseteq \mathcal{L}(\lambda)$. Since Gabriel filters are multiplicatively closed, it follows that $\mathcal{L}^2(\tau) = \mathcal{L} \subseteq \mathcal{L}(\lambda)$. So, $\tau \leq \lambda$, which implies that $\tau = \bigvee_{\sigma \in G} \sigma$. This proves the assertion. $\qquad \square$

(4.6) Corollary. *If R is a left noetherian ring, then $R-\mathbf{sym}$ is a complete sublattice of $R-\mathbf{rad}$.*

Proof. It essentially suffices to show that the meet and join of any family of symmetric radicals in $R-\mathbf{mod}$ is symmetric as well.

For the meet of a family $G \subseteq R-\mathbf{sym}$ this is obvious (even without the noetherian hypothesis), as

$$L \in \mathcal{L}(\bigwedge G) = \bigcap_{\sigma \in G} \mathcal{L}(\sigma)$$

implies that for all $\sigma \in G$ there exists a twosided ideal $I_\sigma \in \mathcal{L}^2(\sigma)$ with $I_\sigma \subseteq L$. Clearly, $I = \sum_{\sigma \in G} I_\sigma \subseteq L$ and

$$I \in \bigcap_{\sigma \in G} \mathcal{L}^2(\sigma) = \mathcal{L}(\bigwedge G).$$

On the other hand, if $L \in \mathcal{L}(\bigvee G)$, then, by the previous result, there exist a finite number of $I_1, \ldots, I_n \in \bigcup_{\sigma \in G} \mathcal{L}(\sigma)$ with $L \supseteq I_1 \ldots I_n$. Since we may choose the I_α to be twosided, $I_1 \ldots I_n \in \mathcal{L}^2(\bigvee G)$, hence $\bigvee G$ is symmetric, indeed. $\qquad \square$

As another application of the construction in (4.4), let us mention:

(4.7) Corollary. *Let R be a left noetherian ring. Then to any prime ideal P of R, we may associate a symmetric radical $\sigma_{R\backslash P}$ whose Gabriel filter $\mathcal{L}(\sigma_{R\backslash P})$ consists of all left ideals L of R which contain a twosided ideal $I \not\subseteq P$.*

(4.8) The previous results actually provide a method to describe essentially *all* symmetric radicals. In order to prove this, we will need the subsets

$$\mathcal{K}(\sigma) = \{P \in Spec(R);\ R/P \in \mathcal{F}_\sigma\}$$

and

$$\mathcal{Z}(\sigma) = \{P \in Spec(R);\ R/P \in \mathcal{T}_\sigma\}$$

of $Spec(R)$. Note already that the set $\mathcal{K}(\sigma)$ resp. $\mathcal{Z}(\sigma)$ is closed under taking generalizations resp. specializations. This means that if $P \in \mathcal{K}(\sigma)$ resp. $P \in \mathcal{Z}(\sigma)$ and if $Q \subseteq P$ resp. $Q \supseteq P$ for some $Q \in Spec(R)$, then $Q \in \mathcal{K}(\sigma)$ resp. $Q \in \mathcal{Z}(\sigma)$ as well.

The sets $\mathcal{K}(\sigma)$ and $\mathcal{Z}(\sigma)$ are linked through the following result:

(4.9) Lemma. *Let σ be a radical in $R-\mathbf{mod}$ and assume σ to be symmetric or R to be left noetherian. If $P \in Spec(R)$, then $P \in \mathcal{Z}(\sigma)$ if and only if $P \notin \mathcal{K}(\sigma)$.*

Proof.   First assume σ to be symmetric and let $\sigma(R/P) \neq 0$, for some $P \in Spec(R)$. Then there exists some $r \in R \setminus P$, such that $\bar{r}$, the class of r in R/P, belongs to $\sigma(R/P)$, i.e., such that $Ir \subseteq P$, for some twosided ideal $I \in \mathcal{L}^2(\sigma)$. But then, since P is prime, $I \subseteq P$ and $P \in \mathcal{Z}(\sigma)$, indeed.

Next, assume R to be left noetherian. If $P \notin \mathcal{K}(\sigma)$, then $0 \neq \sigma(R/P) \subseteq R/P$. Since the ring R/P is prime, it follows from (1.6) that there exists a regular element $\bar{a} \in \sigma(R/P)$. As the regularity of $\bar{a}$ implies that $P = Ann_R^\ell(\bar{a})$, we find that $P \in \mathcal{L}(\sigma)$, which proves the assertion. $\square$

We thus have proved (under the above hypotheses) that

$$\mathcal{Z}(\sigma) = Spec(R) \setminus \mathcal{K}(\sigma).$$

(4.10) Lemma. *Let σ be a symmetric radical and assume R to be left noetherian. Then*

$$\sigma = \bigwedge\{\sigma_{R\setminus P};\ P \in \mathcal{K}(\sigma)\}.$$

Proof. First note that if $P \in \mathcal{K}(\sigma)$ and $I \in \mathcal{L}^2(\sigma)$, then, certainly $I \not\subseteq P$, hence $I \in \mathcal{L}(\sigma_{R\backslash P})$. So, $\sigma \leq \bigwedge_{P \in \mathcal{K}(\sigma)} \sigma_{R\backslash P}$.

On the other hand, pick a twosided ideal

$$I \in \mathcal{L}(\bigwedge_{P \in \mathcal{K}(\sigma)} \sigma_{R\backslash P}) = \bigcap_{P \in \mathcal{K}(\sigma)} \mathcal{L}(\sigma_{R\backslash P}),$$

then $I \not\subseteq P$ for all $P \in \mathcal{K}(\sigma)$. We claim that I then belongs to $\mathcal{L}(\sigma)$, which proves the other inclusion. Indeed, if $I \notin \mathcal{L}(\sigma)$, then we may use the noetherian assumption, to find some twosided ideal $Q \supseteq I$, maximal with respect to the property of not belonging to $\mathcal{L}(\sigma)$. Let us show that Q is a prime ideal, in order to derive a contradiction.

Assume that J and K are twosided R-ideals with the property that $JK \subseteq Q$. We may as well assume that $J, K \supseteq Q$. If both $J \neq Q$ and $K \neq Q$, then $J, K \in \mathcal{L}^2(\sigma)$, by the maximality of Q, hence $JK \in \mathcal{L}^2(\sigma)$. So, $Q \in \mathcal{L}^2(\sigma)$, a contradiction, indeed. $\square$

(4.11) Lemma. *Let K be a generically closed set of prime ideals of a left noetherian R, then*

$$\sigma = \bigwedge\{\sigma_{R\backslash P}; \ P \in K\}$$

is the unique symmetric radical σ in $R-\mathbf{mod}$ with $\mathcal{K}(\sigma) = K$.

Proof. First, note that $\mathcal{K}(\sigma) = K$. Indeed, we have $Q \notin \mathcal{K}(\sigma)$ exactly when $R/Q \in \mathcal{T}_\sigma$, i.e., if and only if $\sigma_{R\backslash P}(R/Q) = R/Q$ for all $P \in K$. But this is equivalent to $Q \not\subseteq P$ for all $P \in K$, i.e., to $Q \notin K$.

Since $\sigma = \bigwedge\{\sigma_{R\backslash P}; \ P \in \mathcal{K}(\sigma)\}$ for any symmetric radical, this proves the assertion. $\square$

Of course, there may be other, non-symmetric radicals τ in $R-\mathbf{mod}$ with $\mathcal{K}(\tau) = \mathcal{K}(\sigma)$. For example, $\mathcal{K}(\sigma_{C(P)}) = \mathcal{K}(\sigma_{R\backslash P})$, for any prime ideal P of a left noetherian ring R.

(4.12) From the foregoing results it thus follows:

(4.13) Corollary. *Let R be a left noetherian ring. There exists a bijective correspondence between generically closed subsets of $Spec(R)$ and symmetric radicals in $R-$mod.*

(4.14) Recall that we have associated to any twosided ideal I of R, which is finitely generated as a left R-ideal a radical σ_I, whose Gabriel filter $\mathcal{L}(\sigma_I) = \mathcal{L}(I)$ consists of all left ideals J of R, with the property that $I^n \subseteq J$, for some positive integer n. By its very definition, it is then clear that the radical σ_I is symmetric. Of course, the definition of σ_I may also be viewed as a special case of the construction given in (4.4).

We may also characterize σ_I by:

(4.15) Proposition. *If I is a twosided ideal of R, which is finitely generated as a left R-ideal, then σ_I is the smallest radical in $R-$mod for which R/I is torsion.*

Proof. Let us call σ the smallest radical in $R-$mod for which R/I is torsion. Clearly, R/I is σ_I-torsion, so $\sigma \leq \sigma_I$. If this inequality is strict, then there exists a nonzero σ_I-torsion left R-module M which is σ-torsionfree. Let $0 \neq m \in M$. Then $Ann_R^\ell(m) \in \mathcal{L}(\sigma_I)$, so there exists a positive integer $n > 1$ such that $I^n m = 0$ and $I^{n-1} m \neq 0$. Choose an element $r \in I^{n-1} \setminus Ann_R^\ell(m)$. Then $Irm \subseteq I^n m = 0$, so the nonzero map

$$R/I \to M : a + I \mapsto arm$$

is left R-linear, contradicting the fact that M is σ-torsionfree. □

(4.16) Proposition. *If I and J are ideals of R which are finitely generated as left R-ideals, then*

(4.16.1) $\sigma_I \vee \sigma_J = \sigma_{IJ}$;

(4.16.2) $\sigma_I \wedge \sigma_J = \sigma_{I+J}$.

Proof. To prove the first assertion, first note that $\sigma_I, \sigma_J \leq \sigma_{IJ}$, hence $\sigma_I \vee \sigma_J \leq \sigma_{IJ}$. Conversely, assume that $M \in R-$mod is torsionfree at $\sigma_I \vee \sigma_J$, then $\sigma_I M = \sigma_J M = 0$. Assume $0 \neq m \in \sigma_{IJ} M$, then we may choose a positive integer $n \geq 1$, minimal with the property that $(IJ)^n m = 0$. So, $(IJ)^{n-1} m \neq 0$, hence $J(IJ)^{n-1} m \neq 0$, as $\sigma_J M = 0$. But then we also have

$$(IJ)^n m = I(J(IJ)^{n-1})m \neq 0,$$

as $\sigma_I M = 0$, as well. This contradiction proves the other inequality $\sigma_I \vee \sigma_J \geq \sigma_{IJ}$.

To prove the second assertion, first note that $\sigma_{I+J} \leq \sigma_I \wedge \sigma_J$. Conversely, if $L \in \mathcal{L}(\sigma_I \wedge \sigma_J) = \mathcal{L}(\sigma_I) \cap \mathcal{L}(\sigma_J)$, then there exist positive integers m and n with the property that $L \supseteq I^n$ resp. $L \supseteq J^m$. With $N = n + m$, we then have $L \supseteq (I + J)^N$, so $L \in \mathcal{L}(\sigma_{I+J})$, which proves the other inequality.   □

If R is left noetherian, then more may be said:

(4.17) Proposition. *If R is a left noetherian ring and if $\{I_\alpha; \ \alpha \in A\}$ is an arbitrary family of twosided ideals of R, then*

$$\bigwedge_{\alpha \in A} \sigma_{I_\alpha} = \sigma_{\sum I_\alpha}.$$

Proof. Let $\sigma = \sigma_{\sum I_\alpha}$. If M is σ-torsion and $0 \neq m \in M$, then there exists a positive integer k such that $(\sum I_\alpha)^k m = 0$. Since $I_\alpha \subseteq \sum I_\alpha$, for each α, this implies that $(I_\alpha)^k m = 0$, i.e., $m \in \sigma_{I_\alpha} M$ for each α. It thus follows that $m \in \bigwedge_{\alpha \in A} \sigma_{I_\alpha} M$, and thus $\sigma \leq \bigwedge_{\alpha \in A} \sigma_{I_\alpha}$.

Conversely, as R is left noetherian, there exist $\alpha_1, \ldots, \alpha_n$ such that $\sum_{i=1}^n I_{\alpha_i} = \sum I_\alpha$. By the previous Lemma, we obtain

$$\sigma = \sigma_{\sum_{i=1}^n I_{\alpha_i}} = \bigwedge_{i=1}^n \sigma_{I_{\alpha_i}} \geq \bigwedge_{\alpha \in A} \sigma_{I_\alpha},$$

which proves the assertion.   □

(4.18) Proposition. *Let R be a left noetherian ring. Then for each prime ideal P of R*

$$\sigma_{R \setminus P} = \bigvee \{\sigma_I; I \nsubseteq P\}.$$

Proof.   Let $\tau = \bigvee\{\sigma_I; I \not\subseteq P\}$. If $I \not\subseteq P$ for some twosided ideal I of R and $L \in \mathcal{L}(\sigma)$, then L contains some power $I^n \not\subseteq P$, i.e., $L \in \mathcal{L}(\sigma_{R\backslash P})$. So, $\sigma_I \leq \sigma_{R\backslash P}$ for all $I \not\subseteq P$, which proves that $\tau \leq \sigma_{R\backslash P}$.

Conversely, if $L \in \mathcal{L}(\sigma_{R\backslash P})$, then $L \subseteq I$ for some twosided ideal I of R with $I \not\subseteq P$. As this implies $L \in \mathcal{L}(I)$ and as $\sigma_I \leq \tau$, it follows that $L \in \mathcal{L}(\tau)$. This shows that $\sigma_{R\backslash P} \leq \tau$, which proves the assertion. $\square$

5. Associated primes

(5.1) *In this section, R will always denote a left noetherian ring.*

A nonzero left R-module N is said to be *prime* if it has the property that $Ann_R^\ell(N) = Ann_R^\ell(L)$ for any nonzero submodule L of N. In this case, $P = Ann_R^\ell(N)$ is a prime ideal of R. Indeed, if $r, s \in R$ are such that $rRs \subseteq P$ and if $s \notin P$, then $0 \neq RsN \subseteq N$. So, $r \in Ann_R^\ell(RsN) = Ann_R^\ell(N) = P$, which proves the assertion.

A prime ideal P of a ring R is said to be *associated* to a left R-module M if there exists a prime submodule N of M (which may then be chosen to be cyclic) such that $P = Ann_R^\ell(N)$. The set of all prime ideals of R which are associated to M will be denoted by $Ass_R(M)$. We will see in a moment that, in the presence of the noetherian hypothesis, if $M \neq 0$ then $Ass_R(M)$ is never empty. This will follow from:

(5.2) Lemma. *Let M be a nonzero left R-module. The maximal elements in the set of annihilators of nonzero submodules of M are prime ideals of R, which are associated to M.*

Proof. The set

$$\mathcal{S} = \{Ann_R^\ell(N);\ 0 \neq N \subseteq M\}$$

is partially ordered by inclusion. Let $P \in \mathcal{S}$ be a maximal element, then there exists $0 \neq N \subseteq M$ such that $P = Ann_R^\ell(N)$. For each $0 \neq L \subseteq N$, clearly $Ann_R^\ell(N) \subseteq Ann_R^\ell(L)$, hence $P = Ann_R^\ell(L)$, by maximality. This proves that N is prime and that $P \in Ass_R(M)$, indeed. $\square$

(5.3) Corollary. *If M is a nonzero left R-module, then $Ass_R(M) \neq \varnothing$.*

Associated primes may sometimes be calculated through dévissage. The main tool here is given by:

(5.4) Lemma. *Assume that*

$$0 \to M' \to M \to M'' \to 0$$

is an exact sequence of left R-modules. Then

$$Ass_R(M') \subseteq Ass_R(M) \subseteq Ass_R(M') \cup Ass_R(M'').$$

Proof. Let $P \in Ass_R(M')$, then there exists a prime submodule $N' \subseteq M'$ such that $P = Ann_R^\ell(N')$ and as $N' \subseteq M$, it follows that $P \in Ass_R(M)$.

Let $P \in Ass_R(M)$, then there exists a prime submodule $N \subseteq M$ such that $P = Ann_R^\ell(N)$. If $N \cap M' \neq 0$, then $N \cap M'$ is a prime submodule of M' and $P \in Ass_R(M')$. If $N \cap M' = 0$, then $N \cong (N + M')/M'$ is isomorphic to a submodule of M'' and it is prime, so $P \in Ass_R(M'')$. □

(5.5) Lemma. *If $\{M_\alpha; \ \alpha \in A\}$ is a family of nonzero left R-modules, then*

$$Ass_R(\bigoplus_{\alpha \in A} M_\alpha) = \bigcup_{\alpha \in A} Ass_R(M_\alpha).$$

Proof. Since $P \in Ass_R(\bigoplus_{\alpha \in A} M_\alpha)$ if and only if $P \in Ass_R(N)$ for some cyclic submodule of $\bigoplus_{\alpha \in A} M_\alpha$, we may assume A to be finite. By induction, we may reduce the problem to the case $M_1 \oplus M_2$ and then the result easily follows from the previous Lemma. □

(5.6) Lemma. *If $M \neq 0$ is a left R-module and N an essential submodule of M, then*

$$Ass_R(N) = Ass_R(M).$$

Proof. As $N \subseteq M$, clearly $Ass_R(N) \subseteq Ass_R(M)$. Let $P \in Ass_R(M)$, then there exists a prime submodule $L \subseteq M$ such that $P = Ann_R^\ell(L)$. Since $0 \neq L \cap N$ is prime, it follows that $P = Ann_R^\ell(L \cap N)$ and $P \in Ass_R(N)$, indeed. □

(5.7) Corollary. *For any nonzero left R-module M, we have*

$$Ass_R(M) = Ass_R(E(M)).$$

(5.8) Corollary. *If M is a nonzero uniform left R-module, then the set $Ass_R(M)$ possesses a single element.*

Proof. Assume that $P, Q \in Ass_R(M)$ with $P \neq Q$, then there exist prime submodules $K, L \subseteq M$ such that $P = Ann_R^\ell(L)$ and $Q = Ann_R^\ell(K)$. Since M is uniform, $L \cap K \neq 0$, hence $P = Ann_R^\ell(L \cap K) = Q$, a contradiction. $\square$

In particular, this applies when M is an indecomposable injective left R-module.

(5.9) Corollary. *Let M be a nonzero finitely generated left R-module. Then the set $Ass_R(M)$ is finite.*

Proof. The injective hull $E(M)$ of M may be decomposed into a finite direct sum

$$E(M) = E_1 \oplus \ldots \oplus E_n$$

of indecomposable injectives. Since

$$Ass_R(M) = Ass_R(E(M)) = \bigcup_{i=1}^{n} Ass_R(E_i)$$

and since each $Ass_R(E_i)$ possesses a single element, this proves the assertion. $\square$

(5.10) Let P be a prime ideal of R. Recall that a nonzero left R-module M is said to be *P-cotertiary*, if $Ass_R(M) = \{P\}$. If M is P-cotertiary for some prime ideal P of R, then we will just say that M is *cotertiary*. In this case, we write $P = ass_R(M)$. A proper submodule N of M with the property that M/N is cotertiary will be called *tertiary* in M. If we want to specify that M/N is P-cotertiary for the prime ideal P of R, then we will say that N is *P-tertiary* in M.

(5.11) Lemma. *Each prime ideal P of R is P-tertiary (in R).*

Proof. Let $Q \in Ass_R(R/P)$, then $Q = Ann_R^{\ell}(L/P)$ for some prime submodule $0 \neq L/P \subseteq R/P$. Clearly, $P(L/P) = 0$, so $P \subseteq Q$. Moreover, $Q(L/P) = 0$, so $QL \subseteq P$, hence $Q \subseteq P$. But then $Q = P$, so $Ass_R(R/P) = \{P\}$, which proves the assertion. $\qquad\square$

If we decompose $E(R/P)$ into a direct sum $E(R/P) = \bigoplus_i E_i$ of indecomposable injective left R-modules, then the previous results show that $Ass_R(E_i) = \{P\}$, i.e., $ass_R(E_i) = P$ for each index i.

In the commutative case, we know, more precisely, that $E(R/P)$ itself is an indecomposable injective R-module for any $P \in Spec(R)$. If R is noncommutative, this is no longer true. However, we have the following result:

(5.12) Proposition. *Let $P \in Spec(R)$ and let $E(R/P) = \bigoplus_i E_i$ be a decomposition into indecomposable injective left R-modules. Then the left R-modules E_i are mutually isomorphic.*

Proof. Let $Q = Q_{cl}(R/P)$. As R/P is essential in Q as a left R/P-module, it is also essential in Q as a left R-module. In particular, $R/P \hookrightarrow E(R/P)$ extends to an injective left R-linear map $Q \hookrightarrow E(R/P)$, which is also essential. The simple ring Q may be decomposed into a direct sum of minimal left ideals, say $Q = \bigoplus_j L_j$, where the L_j are mutually isomorphic.

It thus follows that

$$\bigoplus_i E_i = E(R/P) = \bigoplus_j E(L_j).$$

The Azumaya theorem then yields that for each index i there exists some j such that $E_i = E(L_j)$ (up to isomorphism), so the E_i are isomorphic, indeed. $\qquad\square$

(5.13) Lemma. *Let M be a left R-module. If $\{N_\alpha; \ \alpha \in A\}$ is a family of P-tertiary submodules of M, then $\bigcap_{\alpha \in A} N_\alpha$ is also a P-tertiary submodule of M.*

Proof. The canonical monomorphism

$$M/\bigcap_{\alpha \in A} N_\alpha \hookrightarrow \bigoplus_{\alpha \in A} M/N_\alpha$$

yields that

$$Ass_R(M/\bigcap_{\alpha \in A} N_\alpha) \subseteq \bigcup_{\alpha \in A} Ass_R(M/N_\alpha) = \{P\}.$$

It thus follows that $Ass_R(M/\bigcap_{\alpha \in A} N_\alpha) = \{P\}$, indeed. $\qquad\square$

(5.14) Lemma. *Let M be a left R-module. The following assertions are equivalent:*

(5.14.1) *M is P-cotertiary;*

(5.14.2) *$Ann_M^r(P)$ is an essential submodule of M and P contains any twosided ideal, which annihilates some nonzero submodule of M.*

Proof. Let us first assume M to be P-cotertiary, then P is maximal in the set of annihilators $Ann_R^\ell(N)$ of $0 \neq N \subseteq M$. In particular, P will contain any twosided ideal annihilating such an R-submodule N. On the other hand, to prove that $Ann_M^r(P)$ is essential in M, assume $L \cap Ann_M^r(P) = 0$ for some $0 \neq L \subseteq M$. Equivalently, assume that $Ann_L^r(P) = 0$. However, as $\varnothing \neq Ass_R(L) \subseteq Ass_R(M) = \{P\}$, clearly L is P-cotertiary as well, contradicting $Ann_L^r(P) = 0$!

Conversely, assume (2) and let $N = Ann_M^r(P)$, then we claim that N (which is nonzero as it is essential in M) is prime. Let $0 \neq L \subseteq N$, then $P = Ann_R^\ell(N) \subseteq Ann_R^\ell(L) \subseteq P$ – the last inclusion following from our assumption (2). So, $Ann_R^\ell(L) = Ann_R^\ell(N) = P$. Next, if $Q \in Ass_R(M)$, then $Q = Ann_R^\ell(L)$ for some (prime) $0 \neq L \subseteq M$. Now, our assumption yields that $Q = Ann_R^\ell(L \cap N) = Ann_R^\ell(N) = P$, as $L \cap N \neq 0$, since N is essential in M. This proves that M is P-cotertiary, indeed. $\qquad\square$

(5.15) Let M be a nonzero finitely generated left R-module. A *tertiary decomposition* of a submodule N of M is defined to be a family $\{N_1, \ldots, N_r\}$ of tertiary submodules, with the property that

(5.15.1) $N = N_1 \cap \cdots \cap N_r$;

(5.15.2) the decomposition is irreducible;

(5.15.3) if $ass_R(M/N_i) = P_i$, then $P_j \neq P_j$ for any $j \neq i$.

One of the most important results about tertiary decompositions is given by the next theorem, which we formulate here without proof:

(5.16) Theorem. *(Lesieur, Croisot)* [48] *Let M be a nonzero finitely generated left R-module. Then every submodule N of M possesses a tertiary decomposition. Moreover, if*

$$N = N_1 \cap \cdots \cap N_r = L_1 \cap \cdots \cap L_s$$

are both tertiary decompositions of N in M, then $r = s$ and

$$\{P_1, \ldots, P_r\} = \{Q_1, \ldots, Q_s\}.$$

In addition $Ass_R(M/N) = \{P_1, \ldots, P_r\}$.

The following technical result will be needed in (2.3) in Chapter III:

(5.17) Corollary. *Let M be a finitely generated left R-module. If $N' \subseteq N \subseteq M$ are left R-submodules of M, then there exists some left R-submodule P of M such that $N' = N \cap P$ and $Ass_R(N/N') = Ass_R(M/P)$.*

Proof. If $N' = N$, there is nothing to prove, as we may then take $P = M$. On the other hand, if $N' \neq N$, consider a tertiary decomposition $N' = N_1 \cap \ldots N_n$ of N' in M. If $N_i \not\subseteq N$, then clearly $N_i \cap N$ is tertiary in N, so (after omitting redundant terms) this yields a tertiary decomposition $N' = (N_1 \cap N) \cap \ldots \cap (N_r \cap N)$ of N' in N. Let us now define $P = N_1 \cap \ldots \cap N_r$, then $N' = N \cap P$ and

$$Ass_R(M/P) = \bigcup_{i=1}^{r} Ass_R(M/N_i) = \bigcup_{i=1}^{r} Ass_R(N/N_i \cap N) = Ass_R(N/N'),$$

which proves the assertion. ◻

(5.18) Lemma. *Let M be a nonzero finitely generated R-module. Then there exists a chain*

$$0 = M_0 \subseteq M_1 \subseteq \cdots \subseteq M_n = M$$

with the property that

(5.18.1) M_i *is a tertiary submodule of M_{i+1};*

(5.18.2) M_{i+1}/M_i *is annihilated by its associated prime.*

Proof. We work by induction, i.e., we assume that M_0, M_1, ..., M_r have already been found and that $M_r \neq M$. It then follows that $Ass_R(M/M_r) \neq \varnothing$, hence that there exists some left R-module $M_{r+1} \supset M_r$, such that M_{r+1}/M_r is prime, with $P_{r+1} = Ann_R^\ell(M_{r+1}/M_r)$. This yields a new component M_{r+1} and as long as $M_r \neq M$ this process may be iterated. ◻

Note that $Ass_R(M) \subseteq \{P_1, \ldots, P_n\}$ and that $P_n \ldots P_1 M = 0$.

(5.19) Proposition. *Let M be a nonzero left R-module. For each set of prime ideals $\mathcal{P} \subseteq Ass_R(M)$, there exists a submodule N of M such that $Ass_R(M/N) = Ass_R(M) \setminus \mathcal{P}$ and $Ass_R(N) = \mathcal{P}$.*

Proof. Applying Zorn's Lemma, one finds a left R-submodule $N \subseteq M$ maximal with respect to the property that $Ass_R(N) \subseteq \mathcal{P}$. By (5.4), it thus suffices to show that $Ass_R(M/N) \cap \mathcal{P} = \varnothing$. Pick $P \in Ass_R(M/N)$, then M/N contains some left R-submodule $L/N \cong R/P$. So,

$$Ass_R(L) \subseteq Ass_R(N) \cup Ass_R(L/N) \subseteq \mathcal{P} \cup \{P\},$$

whence $P \notin \mathcal{P}$ by the maximality of N. Now $Ass_R(M/N) \subseteq Ass_R(M) \setminus \mathcal{P}$ and the following chain of inclusions yields the result:

$$Ass_R(M) \subseteq Ass_R(N) \cup Ass_R(M/N)$$
$$\subseteq \mathcal{P} \cup (Ass_R(M) \setminus \mathcal{P}) = Ass_R(M).$$

◻

(**5.20**)   There exists another generalization of primary decomposition to the noncommutative case. Indeed, recall that a P-cotertiary left R-module M is P-*coprimary* if there exists some positive integer n with $P^n M = 0$. A submodule N of M is P-*primary* in M if M/N is P-coprimary. If the prime ideal P is irrelevant, we just say that N is *primary* in M. If I is a left ideal we define $rad(I)$ as the intersection of all prime ideals containing $Ann_R^\ell(R/I) = (I : R)$.

With these definitions we have:

(**5.21**) **Proposition.** *A left ideal I of the ring R is primary in R if and only if for any $a, b \in R$ with $aRb \subseteq I$ and $b \notin I$, we have $a \in rad(I)$.*

Proof.   If I is primary and $aRb \subseteq I$ for $b \notin I$, then $a(Rb + I)/I = 0$, so $a \in ass_R(R/I) = P$. As $P^n(R/I) = 0$ for some positive integer n, then $P^n \subseteq I$ and $P \subseteq rad(I)$, this proves that $a \in rad(I)$.

Conversely, let $P \in Ass_R(R/I)$ and assume I satisfies our assumption. There exists some $b \notin I$ with

$$P = \{a \in R; \; aRb \subseteq I\} = Ann_R^\ell(Rb + I/I),$$

so $P \subseteq rad(I)$. However, clearly $P \supseteq rad(I)$ as well, so $P = rad(I)$. It follows that I is tertiary in R, hence also primary in R, as $rad(I)^n \subseteq I$ for some positive integer n.   □

It thus follows, in particular, that $Ass_R(R/I) = \{rad(I)\}$ for any primary left ideal I of R.

(**5.22**)   Primary decomposition may now be defined in a similar way as tertiary decomposition. However, in general, primary decomposition in this sense does not exist. In fact, it is rather easy to verify that a sufficient condition for a primary decomposition of $M \in R-\textbf{mod}$ to exist is that every irreducible left R-submodule of M be primary, a condition which is obviously met in the commutative case.

It thus appears that a complete theory of primary decomposition in the previous sense is bound to be rather unsatisfactory for general noncommutative rings. The reason why we mentioned primary decomposition (and primary submodules) at all, resides in the fact that primary and tertiary submodules of M are tightly connected, when M satisfies the Artin-Rees condition, as we will see in Chapter III below.

Let us finish this section with a brief study of the relationship between (symmetric) radicals and associated primes.

We start with:

(5.23) Proposition. *Let M be a left R-module and σ a symmetric radical in $R-\mathbf{mod}$. Then:*

(5.23.1) *if $M \in \mathcal{T}_\sigma$, then $Ass_R(M) \subseteq \mathcal{Z}(\sigma)$;*

(5.23.2) *if $Ass_R(M) \subseteq \mathcal{K}(\sigma)$, then $M \in \mathcal{F}_\sigma$.*

Proof. To prove the first assertion, assume $P \in Ass_R(M)$. Then there exists a (prime) submodule $N \subseteq M$ such that $P = Ann_R^\ell(N)$ and by definition, we may then assume that $N = Rm$. Since $N \in \mathcal{T}_\sigma$, there exists some $I \in \mathcal{L}^2(\sigma)$, with $Im = 0$. So $I \subseteq P$, hence $P \in \mathcal{Z}(\sigma)$, indeed.

For the second assertion, if M is not σ-torsionfree, i.e. $\sigma M \neq 0$, then

$$\varnothing \neq Ass_R(\sigma M) \subseteq Ass_R(M) \subseteq \mathcal{K}(\sigma).$$

This is a contradiction, as $\sigma M \in \mathcal{T}_\sigma$ implies that $Ass_R(\sigma M) \subseteq \mathcal{Z}(\sigma)$, by the first statement. $\square$

The following result (for symmetric radicals) will be strengthened considerably in (2.16) in Chapter III:

(5.24) Lemma. *Let σ be a symmetric radical. Then the following assertions are equivalent:*

(5.24.1) *σ is stable;*

(5.24.2) *the torsion class of σ is given by*

$$\mathcal{T}_\sigma = \{M \in R\text{-}\mathbf{mod}; \; Ass_R(M) \subseteq \mathcal{Z}(\sigma)\}.$$

Proof. To prove that (2) implies (1), choose $M \in \mathcal{T}_\sigma$. Then

$$Ass_R(E(M)) = Ass_R(M) \subseteq \mathcal{Z}(\sigma),$$

hence we also have $E(M) \in \mathcal{T}_\sigma$. So σ is stable, indeed.

Conversely, assume that σ is stable and let us prove (2). First note that

$$\mathcal{T}_\sigma \subseteq \{M \in R\text{-}\mathbf{mod}; \; Ass_R(M) \subseteq \mathcal{Z}(\sigma)\},$$

by (5.23). On the other hand, choose a left R-module M with the property that $Ass_R(M) \subseteq \mathcal{Z}(\sigma)$, and let us show that $M \in \mathcal{T}_\sigma$. Since $M \in \mathcal{T}_\sigma$ exactly when $N \in \mathcal{T}_\sigma$, for any finitely generated submodule $N \subseteq M$, we may as well assume M to be finitely generated. By (5.16) there exists a tertiary decomposition of 0 in M, i.e., a family of tertiary submodules $\{N_1, \ldots, N_r\}$ with the property that $N_1 \cap \cdots \cap N_r = 0$, that $Ass_R(M/N_i) = \{P_i\}$ and that

$$Ass_R(M) = \{P_1, \ldots, P_r\} \subseteq \mathcal{Z}(\sigma).$$

As there exists a monomorphism $M \hookrightarrow \bigoplus_{i=1}^r M/N_i$, we may assume that M is P-cotertiary and that $P \in \mathcal{Z}(\sigma)$. By (5.14), $Ann_M^r(P)$ is an essential submodule of M and by the stability of σ, it suffices to show that $Ann_M^r(P) \in \mathcal{T}_\sigma$. Let $m \in Ann_M^r(P)$, then $Pm = 0$ and hence there exists a surjective map

$$R/P \to R/Ann_R^\ell(m).$$

As $P \in \mathcal{Z}(\sigma)$, we obtain that $Ann_R^\ell(m) \in \mathcal{L}(\sigma)$, so $Rm \in \mathcal{T}_\sigma$, whence $Ann_M^r(P) \in \mathcal{T}_\sigma$, indeed. $\square$

(5.25) Lemma. *Let σ be a stable symmetric radical. For any left R-module M, we have:*

(5.25.1) $Ass_R(\sigma M) = Ass_R(M) \cap \mathcal{Z}(\sigma)$;

(5.25.2) $Ass_R(M/\sigma M) = Ass_R(M) \cap \mathcal{K}(\sigma)$;

(5.25.3) $Ass_R(M) = Ass_R(\sigma M) \cup Ass_R(M/\sigma M)$.

Proof. Let $\mathcal{P} = Ass_R(M) \cap \mathcal{Z}(\sigma)$. By (5.19), there exists a submodule N of M such that

$$Ass_R(N) = Ass_R(M) \cap \mathcal{Z}(\sigma)$$

and

$$Ass_R(M/N) = Ass_R(M) \cap \mathcal{K}(\sigma).$$

Clearly

$$Ass_R(M) = Ass_R(N) \cup Ass_R(M/N).$$

Now, since $Ass_R(M/N) \subseteq \mathcal{K}(\sigma)$, we have that $M/N \in \mathcal{F}_\sigma$ and as $Ass_R(N) \subseteq \mathcal{Z}(\sigma)$, we obtain that $N \in \mathcal{T}_\sigma$. So, $N = \sigma M$, indeed. $\square$

From this, it now easily follows:

(5.26) Corollary. *Let σ be a stable symmetric radical. Then:*

(5.26.1) $\mathcal{T}_\sigma = \{M \in R-\textbf{mod}; \ Ass_R(M) \subseteq \mathcal{Z}(\sigma)\}$;

(5.26.2) $\mathcal{F}_\sigma = \{M \in R-\textbf{mod}; \ Ass_R(M) \subseteq \mathcal{K}(\sigma)\}$.

6. The second layer condition

In this section, we collect some results concerning the second layer condition, whose proofs we have included for the reader's convenience. (For full details we refer to the literature, cf. [9, 37, 52, et al]). Note that our main aim will be to give a proof of (6.5) below.

(6.1) Recall from [37, 52] for example, that a left R-module M is said to be *uniform* if it is essential over any of its nonzero left R-submodules, i.e., if N and N' are submodules of M with $N \cap N' = 0$, then $N = 0$ or $N' = 0$. It is easy to see that this is equivalent to asserting that the injective hull $E(M)$ of M is an indecomposable injective (i.e., if E is a nonzero injective left R-submodule of $E(M)$, then $E = E(M)$). In R is a commutative noetherian ring, it is thus clear that $E(M) = E(R/P)$ for some (uniquely determined) prime ideal P of R.

We say that $P \in Spec(R)$ satisfies the *left second layer condition*, if there does *not* exist a prime ideal $Q \subset P$ and a short exact sequence of finitely generated uniform left R-modules

$$0 \to L \to M \to N \to 0,$$

with the property that

(6.1.1) $L = Ann_M^r(P)$;

(6.1.2) L is torsionfree as a left R/P-module;

(6.1.3) $Q = Ann_R^\ell(N) = Ann_R^\ell(N')$ for any $0 \neq N' \subseteq N$.

If we may weaken the second condition to:

(6.1.4) $Ann_R^\ell(L') = P$ for all left R-submodules $0 \neq L' \subseteq L$,

then we will say that P satisfies the *strong left second layer condition*.

The right analogues of these notions are defined similarly. If every prime ideal in $X \subseteq Spec(R)$ resp. $Spec(R)$ satisfies the (strong) left second layer condition, then we say that X resp. R satisfies the (strong) second layer condition.

Let us start with Jategaonkar's Main Lemma (see also [9, 4.1], [37, 11.1], [52, 4.3.10.]):

(6.2) Proposition. *(Jategaonkar's Main Lemma) Let R be a left noetherian ring and let $P, Q \in Spec(R)$. Consider an exact sequence of finitely generated uniform left R-modules*

$$0 \to L \to M \to N \to 0$$

and put $Ann_R^\ell(M) = A$. Assume the following assertions to hold:

(6.2.1) $P = Ann_R^\ell(L)$, $L = Ann_M^r(P)$ *and L is torsionfree as an R/P-module;*

(6.2.2) *for any submodule M' of M, we have $Ann_R^\ell(M') = A$ or $M' \subseteq L$;*

(6.2.3) $Q = Ann_R^\ell(N)$.

Then $PQ \subseteq A \subseteq P \cap Q$ and

*(**a**) if $A \subset P \cap Q$, then N is torsionfree as an R/Q-module and $P \rightsquigarrow Q$;*

*(**b**) if $A = P \cap Q$, then N is not torsionfree as an R/Q-module and $A = Q \subset P$.*

*(Note that (**b**) cannot occur if P satisfies the left second layer condition).*

Proof. It is clear that $PQ \subseteq A \subseteq P \cap Q$. Indeed, since $QN = 0$, obviously $QM \subseteq L$. Hence $PQM \subseteq PL = 0$, so $PQ \subseteq A = Ann_R^\ell(M)$. On the other hand, $AM = 0$ implies that $AL = 0$ and thus $A \subseteq P$. Since we also have $AN = 0$, as $AM = 0$, whence $A \subseteq Q$, this yields the assertion.

(**a**) Let us now first assume A to be strictly contained in $P \cap Q$. Consider an arbitrary twosided ideal I of R with $A \subset I \subseteq P \cap Q$ (e.g., take $I = P \cap Q$), then we claim that $Ann_R^\ell(I/A) = P$ and $Ann_R^r(I/A) = Q$, which easily yields that $P \rightsquigarrow Q$, of course.

For the first assertion, put $U = Ann_R^\ell(I/A)$. Then $UI \subseteq A$, hence $UIM = 0$. As $A \subset I$, it follows that $0 \neq IM \subseteq L$, so

$$U \subseteq Ann_R^\ell(IM) = Ann_R^\ell(L) = P.$$

(For the first equality, note that if $Ann_R^\ell(IM) = J$ would strictly contain P, then the essentiality of J/P in R/P would imply the existence of some $c \in \mathcal{C}_R(P) \cap P$. Since $c(IM) = 0$, this implies $IM \subseteq \sigma_L = 0$, a contradiction!) As $PI \subseteq PQ \subseteq A$, we have $P \subseteq U$ as well, so $P = U$, indeed.

For the second assertion, put $V = Ann_R^r(I/A)$. Then $V = Ann_R^r(I/A)$, hence $IV \subseteq A$ and $IVM = 0$. Since $A \subset I \subseteq Ann_R^\ell(VM)$, condition (2) implies that $VM \subseteq L$, so $V \subseteq Ann_R^\ell(N) = Q$. Conversely, as $PQ \subseteq A$ and $I \subseteq P$, we obtain $IQ \subseteq A$. So $Q \subseteq V$ as well, which yields that $Q = V$, as claimed.

Let us now suppose, for a moment, that N is not torsionfree as an R/Q-module. Then we may pick $x \in M \setminus L$ and $d \in \mathcal{C}_R(Q)$, with $dx \in L$. As d acts regularly on $P \cap Q/A$, multiplying (on the right) by d induces an isomorphism between $P \cap Q/A$ and $(P \cap Q)d + A/A$. Localizing the exact sequence of left R/P-modules

$$0 \rightarrow (P \cap Q)d + A/A \rightarrow P \cap Q/A \rightarrow P \cap Q/(P \cap Q)d + A \rightarrow 0$$

at the regular elements of R/P then easily shows the R/P-module $P \cap Q/(P \cap Q)d + A$ to be torsion. Since $(P \cap Q)x$ is the image of $P \cap Q \subseteq R$ under the morphism $\phi : R \rightarrow M$, which maps $r \in R$ to $rx \in M$ and since, clearly, $(P \cap Q)d + A \subseteq Ker(\phi)$, it follows that $(P \cap Q)x$ is a quotient of $P \cap Q/(P \cap Q)d + A$. We thus obtain that $(P \cap Q)x$ is torsion as an R/P-module.

On the other hand, since $x \in M$, we obtain that $P(P \cap Q)x \subseteq Ax = 0$, so $(P \cap Q)x \subseteq Ann_M^r(P) = L$. But, as L is assumed to be torsionfree at P and since we just verified $(P \cap Q)x$ to be torsion, it follows that $(P \cap Q)Rx = (P \cap Q)x = 0$. We thus find that $Ann_R^\ell(Rx) \supseteq P \cap Q \supset A$. From (2), it follows that $Rx \subseteq L$, hence $x \in L$ – a contradiction.

(b) Let us now assume that $A = P \cap Q$. Then $QP \subseteq A$, hence $QPM = 0$. Since $PM \neq 0$ (as $Ann^r_M(P) = L \neq M$) and since M is uniform, we have $PM \cap L \neq 0$. As $Q(PM \cap L) = 0$, it follows from (1) that $Q \subseteq P$, hence that $A = Q$. Moreover, $PM \neq 0$ yields that $A \neq P$, hence that $Q \subset P$.

To conclude, assume for a moment N to be torsionfree as an R/Q-module, then N is isomorphic to a uniform left ideal of R/Q. Using [37, Corollary 6.25.], it follows that there exists some positive integer n (the *uniform dimension* of R/Q), such that N^n contains an isomorphic copy of R/Q, say with inverse image K in M^n. The R/Q-linear map $K \to R/Q$ then splits, as R/Q is free, so $K = R/Q \oplus (L^n \cap K)$ (up to isomorphism). On the other hand, L is essential in M, so $L^n \cap K$ is essential in K – a contradiction. This proves that N cannot be torsionfree as a left R/Q-module, which finishes the proof. $\square$

Recall the following classical result:

(6.3) Proposition. [52, 4.3.13] *Let R be a noetherian ring and let $X \subseteq Spec(R)$ be a clique. The following assertions are equivalent:*

(6.3.1) *X satisfies the left second layer condition;*

(6.3.2) *if $P \in X$ and if M is a finitely generated left R-submodule of $E(R/P)$, then M possesses a chain of submodules*

$$0 = M_0 \subseteq \ldots \subseteq M_n = M,$$

such that M_i/M_{i-1} is isomorphic to a uniform left ideal of R/P_i for some $P_i \in X$;

(6.3.3) *for each $P \in X$, the only prime ideal which is the annihilator of a finitely generated left R-submodule of $E(R/P)$ is P itself.*

Proof. To prove that (1) implies (2), let us prove the conclusion for any finitely generated submodule $M \subseteq E_R(L_1) \oplus \ldots \oplus E_R(L_t)$, where L_i is a uniform left ideal of R/P_i with $P_i \in X$. Given such a module

M, the exact sequence

$$0 \to E_R(L_t) \to E_R(L_1) \oplus \ldots \oplus E_R(L_t) \to E_R(L_1) \oplus \ldots \oplus E_R(L_{t-1}) \to 0$$

yields an exact sequence

$$0 \to M \cap E_R(L_t) \to M \to M' \to 0$$

with $M' \hookrightarrow E_R(L_1) \oplus \ldots \oplus E_R(L_{t-1})$. Induction on t thus reduces the problem to the case $t = 1$.

In this situation, we write $L = L_1$, $P = P_1$ and, up to replacing L by $L \cap M$, we may suppose $L \subseteq M$. Note that $Ann_M^r(P)$ is a torsionfree uniform R/P-module, hence it is isomorphic to a left ideal of R/P. We may thus assume that $L = Ann_R^r(P)$.

If $L = M$, we have nothing to prove. Otherwise, let N be a uniform submodule of M/L with inverse image K in M. By [52, 4.3.9] there exists a submodule K' of K with the property that if $Q = ass_R(K')$ and $U = Ann_{K'}^l(Q)$, then $U = L \cap K'$. The second layer condition applied to K' yields that K'/U is isomorphic to R/P' for some $P' \in X$. Note also that $K'/U \hookrightarrow N$.

Finally, let us note that M/L is an essential extension of some direct sum of uniform submodules $N_1, \ldots, N_s$. It thus follows that $M/L \subseteq E_R(N_1') \oplus \ldots \oplus E_R(N_s')$, for some uniform left ideals N_j' of R/P_j, with $P_j \in X$. Applying noetherian induction to M, then yields that M/L possesses a composition series of the desired type.

The previous argument also shows that (3) implies (1). Indeed, let M be a finitely generated submodule of $E_R(R/P)$ with $P \neq A = Ann_R^l(M) \in Spec(R)$. Clearly, one of the prime ideals P_i provided by (2) must be equal to A and must arise, in the induction process, as the annihilator of the top piece of an M'. But then $Ann_R^l(M') = A$, contradicting the second layer condition applied to M'.

To prove that (2) implies (1), consider data such as in the previous Proposition and assume the second layer condition to fail for M. So,

$Ann_R^\ell(M) = Q \subset P$ and N is torsion as an R/Q-module. By assumption, since $M \hookrightarrow E_R(L)$, there exists a chain

$$0 = M_0 \subseteq M_1 \subseteq \ldots \subseteq M_n = M,$$

with M_i/M_{i-1} isomorphic to a uniform left ideal of R/P_i for some $P_i \in X$. As M is uniform, $M_1 \cap L \neq 0$, so $P = P_1$ and $M_1 \subseteq L = Ann_M^r(P)$. If $M_2 \subseteq L$, we may eliminate M_1 from the chain. Hence, without loss of generality, we may assume that $M_2 \not\subseteq L$ and consider $(M_2 \cap L)/M_1$. If this is nonzero, it is isomorphic to a uniform left ideal of R/P_2 and, being a submodule of L/M_2, it is annihilated by P. Hence $P \subseteq P_2$, so $P^2 M_2 = 0$. As $Ann_N^r(P) = 0$, this implies $M_2 \subseteq L$, contradicting our hypotheses.

This proves that $(M_2 + L)/L$ is isomorphic to M_2/M_1, hence

$$P_2 = Ann_R^\ell(M_2/M_1) = Ann_R^\ell((M_2 + L)/L) = Ann_R^\ell(N) = Q.$$

Moreover, M_2/M_1 is isomorphic to a left ideal of R/P_2 and $(M_2 + L)/L$ is torsion as an R/Q-module – a contradiction!

Finally, that (3) also implies (1) easily follows from the fact that if, as before, the second layer condition fails for M, then $Ann_R^\ell(M) = Q \subset P$, a contradiction. This finishes the proof. $\square$

A more precise version of the previous result may now be given as follows:

(6.4) Proposition. *Let R be a noetherian ring and X a clique of R. Consider the following assertions:*

(6.4.1) *X satisfies the strong left second layer condition;*

(6.4.2) *if $P \in X$ and M is a finitely generated and uniform left R-module such that $ass_R(M) = P$, then M possesses a chain of submodules*

$$0 = M_0 \subseteq \ldots \subseteq M_n = M,$$

such that $ass_R(M_i/M_{i-1}) = Ann_R^\ell(M_i/M_{i-1}) \in X;$

(6.4.3) if $P \in X$ and M is a finitely generated P-cotertiary left R-module, then M possesses a chain of submodules

$$0 = M_0 \subseteq \ldots \subseteq M_n = M,$$

such that

$$ass_R(M_i/M_{i-1}) = Ann_R^\ell(M_i/M_{i-1}) = P_i \in X$$

and $Ann_{M/M_{i-1}}^r(P_i) = M_i/M_{i-1}$;

(6.4.4) if $P \in X$ and M is a finitely generated and uniform left R-module with $ass_R(M) \in X$ and $Ann_R^\ell(M) \in Spec(R)$, then M is a prime module.

Then the following implications hold: $(4) \Longleftrightarrow (1) \Longrightarrow (2) \Longleftrightarrow (3)$.

Proof. Let us first show that (1) implies (2). Let $\mathcal{S}$ be the set of submodules N of M for which there exists a chain ending in N and whose factors are as in the statement. Let N be a maximal element of this set and Q an element of $Ass_R(M/N)$. Pick $M' \subseteq M$ such that $Ann_R^\ell(M' + N/N) = ass_R(M' + N/N) = Q$ and with $Ann_R^\ell(M')$ maximal in the set

$$\{Ann_R^\ell(Y); \; Y \subseteq M, Y \not\subseteq N\}.$$

It obviously follows that the requirements in (6.2) are fulfilled by the exact sequence

$$0 \to M' \cap N \to M' \to M'/M' \cap N \to 0,$$

hence, either $Q \subset P$ or $P \rightsquigarrow Q$. The first case cannot hold since this would contradict the strong second layer condition on $P \in X$. It follows that $N \subset M' + N \subseteq M'' \in \mathcal{S}$, where $M''/N = Ann_{M/N}^\ell(Q)$, thus contradicting the maximality of N, hence (2) holds, indeed.

To prove that (2) implies (3), we decompose the injective hull of M as a direct sum $E(M) = \bigoplus_{i=1}^n E_i$ of indecomposable injectives and take $M_i = (\bigoplus_{j=1}^{i} E_j) \cap M$. We have $ass_R(E_i) = ass_R(M) = P \in X$ and M_i/M_{i-1} obviously embeds into E_i, hence (2) may be applied to M_i/M_{i-1}. Joining the chains thus obtained, this yields the desired conclusion.

Clearly (3) implies (2) and (4) implies (1), hence there only remains to be shown that (1) implies (4). To this end, consider a finitely generated uniform left R-module M with $Ann_R^\ell(M) = Q \in Spec(R)$ and $ass_R(M) = P \supseteq Q$. If we consider a chain as in (2) we have that $P_n \dots P_1 \subseteq Q$ and $Q \subseteq P_i$ for every i. Hence there exists i such that $P_i = Q$, so P and Q belong to X. By [45, 8.2.8], we have that $dim(R/P) = dim(R/Q)$, where $dim(-)$ denotes the classical Krull dimension, hence $P = Q$. $\qquad\Box$

Finally, let us note:

(6.5) Corollary. *(cf. [9, 4.3], [37, 11.4]) Let M be a finitely generated left R-module and let*

$$X = \bigcup\{cl^\ell(P); \; P \in Ass_R(M)\}.$$

If X satisfies the strong left second layer condition, then there exists a chain

$$0 = M_0 \subseteq M_1 \subseteq \dots \subseteq M_n = M,$$

such that for each $1 \leq i \leq n$ the left R-module M_i/M_{i-1} is uniform and has the property that $Ann_R^\ell(M_i/M_{i-1}) = P_i \in X$. In particular, it follows that $P_1 \dots P_n M = 0$.

(6.6) Proposition. *Let P and Q be prime ideals of a noetherian ring R. The following assertions are equivalent:*

(6.6.1) *there exists a link $P \rightsquigarrow Q$;*

(6.6.2) *there exists an exact sequence*

$$0 \to L \to M \to N \to 0$$

of finitely generated uniform left R-modules, such that L is isomorphic to a left ideal of R/P and N is isomorphic to a left ideal of R/Q.

Proof.   If an exact sequence as in (2) exists, all submodules of N are torsionfree as R/Q-modules. So, case (a) in (6.2) cannot occur and $P \rightsquigarrow Q$, indeed.

To prove that (1) implies (2), assume that $P \rightsquigarrow Q$ through some twosided ideal A. Without loss of generality, we may assume $A = 0$, hence $0 \neq P \cap Q$ is torsionfree, both as a left R/P-module and a right R/Q-module. Since $Ann_R^r(P \cap Q) = Q$, clearly $Ann_R^r(Q) \subseteq Q$ and $Ann_R^r(P) = Q$ (as $PQ = 0$).

If L is a nonzero left ideal of R, then either $QL = 0$ and then $L \subseteq Ann_R^r(Q) \subseteq Q$ or $0 \neq QL \subseteq L \cap Q$. It thus follows that Q is essential as a left ideal of R. Now, since both $P \cap Q$ and $Q/P \cap Q$ are torsionfree left R/P-modules (the latter as it is isomorphic to an ideal of R/P), we find that Q is a torsionfree left R/P-module as well.

Let E be the injective hull of R in $R-\mathbf{mod}$ and put $K = Ann_R^\ell(P)$. It is easy to see that $K \cap R = Ann_R^r(P) = Q$, so

$$R/Q = R/K \cap R \cong R + K/K \hookrightarrow E/K.$$

Decompose E into a finite direct sum $E_1 \oplus \ldots \oplus E_n$ of uniform (indecomposable) injectives, then $K = (K \cap E_1) \oplus \ldots \oplus (K \cap E_n)$, so

$$E/K = E_1/K \cap E_1 \oplus \ldots \oplus E_n/K \cap E_n.$$

It follows that some component $E_j/K \cap E_j$ contains a copy of a uniform left ideal U of R/Q. Moreover, K is an essential submodule of E, which is torsionfree as a left R/P-module, being an essential extension of Q (which is torsionfree, as we have just seen).

Finally, consider a finitely generated submodule M' of E_j, such that $K \cap E_j \subseteq M' \subseteq E_j$ and $M'/K \cap E_j \cong U$. Let M be a finitely generated submodule of M', which is not contained in K. If we put $L = K \cap M =$

$Ann^r_M(P)$ and $N = M/L \neq 0$, we obtain the required exact sequence, as $N \hookrightarrow M'/K \cap E_j$. □

One now also has (cf. [9, 4.3]):

(6.7) Proposition. *Let M be a finitely generated left R-module and let*

$$X = \bigcup \{cl^\ell(P); \ P \in Ass_R(M)\}.$$

If R is noetherian, X satisfies the left second layer condition and $Ann^r_M(P)$ is torsionfree[3] left R/P-module for every $P \in Ass_R(M)$, then there exists a chain

$$0 = M_0 \subseteq M_1 \subseteq \ldots \subseteq M_n = M$$

such that for each $1 \leq i \leq n$ the left R-module M_i/M_{i-1} is uniform and has the property that $Ann^\ell_R(M_i/M_{i-1}) = P_i \in X$. (Of course, again, it follows that $P_1 \ldots P_n M = 0$).

Proof. By induction, we may assume the result to hold for any proper factor of M. Let $P \in Ass_R(M)$ and let N be a maximal uniform left R-submodule of $L = Ann^r_M(P) \neq 0$. We want to show that $Ass_R(M/N) \subseteq X$ and that if $Q \in Ass_R(M/N)$, then $Ann^r_{M/N}(Q)$ is torsionfree as a left R/Q-module, for then our induction assumption finishes the proof.

So, let U be a uniform left R-submodule of M/N with $Q = Ann^\ell_R(U) = ass_R(U)$. Write U as V/N for some left R-submodule $N \subset V \subseteq M$.

If V is uniform, then one may shrink V if necessary and apply (6.2) to derive that $P \rightsquigarrow Q$. Moreover, in the presence of the noetherian hypothesis, it follows from Jategaonkar's Main Lemma, that some submodule of U and hence U itself is torsionfree as a left R/Q-module.

If V is not uniform, then we may find some uniform left R-submodule

---

[3]In the terminology of [45], this just says that M is *tame*

of M embedded into U. It follows that $Q \in Ass_R(M)$ and our assumptions imply U to be torsionfree over R/Q, indeed.   □

Localizability and classical localizability are usually verified by checking the so-called intersection property. Indeed, recall that if $X \subseteq Spec(R)$ is a set of prime ideals, then one says that X satisfies the *weak left intersection property*, if for every left ideal L of R with $L \cap C_R(P)r \neq \varnothing$ for any $P \in X$ and any $r \in R$, we have $L \cap C_R(X) \neq \varnothing$. Clearly, if X is finite, then X satisfies the weak left intersection property.

(6.8) Proposition. *Assume R to be noetherian and X to be a subset of $Spec(R)$. If X satisfies the weak left intersection property and the left second layer condition, then $C_R(X)$ is a left denominator set in R.*

Proof. (cf. [9]) Assume $C = C_R(X)$ does not satisfy the (first) left Ore condition, then there exists some $c \in C$ with the property that R/Rc is not C-torsion. Since R/Rc is left noetherian, there exists a quotient K of R/Rc with the property that K is not C-torsion, but every proper quotient of K is. Clearly K is uniform and generated by a C-torsion generator. The weak intersection condition then implies that we may find some prime ideal $P \in X$ with the property that K is not $C_R(P)$-torsion.

Let us put $K' = Ann_K^r(P)$, then we claim that $K' = K$. This will then finish the proof. Indeed, we then have $PK = 0$, so K is a left R/P-module, which is not torsion, but generated by a torsion element – a contradiction.

So, suppose $K' \subset K$. As K' is either zero or a uniform R/P-module, obviously K' is either torsionfree or torsion at $C_R(P)$. Suppose for a moment $x \in K'$ not to be $C_R(P)$-torsion. Then $Ann_R^\ell(x) + P/P$ is not essential in R/P, implying that $R/Ann_R^\ell(x) + P$ is not torsion as a left R/P-module.   However, $R/Ann_R^\ell(x) + P$ is isomorphic to Rx/Px, so K/Px is not $C_R(P)$-torsion, as it contains Rx/Px. Unless $Px = 0$ (or, equivalently, $x \in K'$), this leads to a contradiction. We thus obtain that $0 \neq K'$ is $\mathcal{C}_R(P)$-torsionfree.

To conclude, shrink K to a submodule M containing K' and such that Jategaonkar's Main Lemma applies to the short exact sequence

$$0 \to K' \to M \to M/K' \to 0$$

and put $Q = Ann_R^\ell(M/K') = ass_R(M/K')$. The Main Lemma, combined with the fact that $P \in X$ satisfies the left second layer condition, now implies that $P \rightsquigarrow Q$, through the twosided ideal $Ann_R^\ell(M) \subset P \cap Q$ and the left R-module M/K' is $\mathcal{C}_R(Q)$-torsionfree. Since X is left link closed, we find $Q \in X$, so M/K' is $\mathcal{C}_R(X)$-torsionfree. Hence K/K' is not $\mathcal{C}_R(X)$-torsion and this contradicts the fact that $K' \neq 0$. This proves the assertion. $\square$

If $X \subseteq Spec(R)$ is a set of prime ideals, then we say that X satisfies the *incomparability property* if any two prime ideals in X are incomparable, i.e., if $P \subseteq Q$ belong to X, then $P = Q$.

One of the main results about classical localizability may then be formulated as:

(6.9) Theorem. [9, 37, 52, et al] *Let R be a (noetherian) ring and $X \subseteq Spec(R)$ a set of prime ideals of R. Then X is classically left localizable if and only if:*

(6.9.1) *X is left link closed;*

(6.9.2) *X satisfies the left second layer condition;*

(6.9.3) *X satisfies the weak left intersection property;*

(6.9.4) *X satisfies the incomparability property.*

Proof. (The proof below is the one given in [9]) To prove that the conditions imply the classical left localizability of X, first note that we already know (from the previous result) that $R_X = C^{-1}R$ to exist, for $C = \mathcal{C}_R(X)$.

Let $P \subseteq I$ be a left ideal of R with the property that I/P is essential in R/P. Then $I \cap \mathcal{C}_R(P)r \neq \varnothing$ for every $r \in R$, as $r + I$ belongs to the torsion module R/I. Consider $P \neq Q \in X$. By the incomparability assumption, $(P + Q)/Q$ is a nonzero twosided ideal of R/Q, hence it is essential as a left ideal. So, $P \cap \mathcal{C}_R(Q)r \neq \varnothing$ for all $r \in R$, i.e., we also have $I \cap \mathcal{C}_R(Q)r \neq \varnothing$. The weak intersection property thus implies that $I \cap C \neq \varnothing$. This shows that any essential left ideal of $\overline{R} = R/P$ intersects $\overline{C} = \{c + P;\ c \in C\}$, so $\overline{C}^{-1}\overline{R} = Q_{cl}(\overline{R})$. Hence $R_X/R_X P \cong \overline{C}^{-1}\overline{R}$ is artinian.

Assume K to be a simple left R_X-module. Then K is a C-torsionfree uniform left R-module and, moreover, every proper quotient of K is C-torsion.

Let us show that $PK = 0$ for some $P \in X$. By the weak intersection property, there exists some $P \in X$ such that K is not torsion at P. Let

$$L = \{x \in K;\ Px = 0\},$$

then we want to show that $L = K$, thus proving our claim. So, assume L to be strictly contained in K. As L is either zero or uniform as a left R/P-module, it is either σ_P-torsionfree or σ_P-torsion. Assume that $x \in K$ is not torsion at P, then $(Ann_R(x) + P)/P$ is an essential left ideal of R/P, so $R/(Ann_R(x) + P)$ is not torsion as a left R/P-module. As $R/(Ann_R(x) + P) \cong Rx/Px$, this implies K/Px not to be torsion at P – a contradiction, unless $Px = 0$ or, equivalently, $x \in L$. This proves that $L \neq 0$ and that L is σ_P-torsionfree.

Let us now shrink K to a submodule M containing L and such that the Main Lemma applies to the exact sequence

$$0 \to L \to M \to M/L \to 0,$$

and let $Q = Ann_R^\ell(M/L) = ass_R(M/L)$. By Jategaonkar's Main Lemma and the fact that P satisfies the left second layer condition, we see that $P \rightsquigarrow Q$ through $Ann_R^\ell(M) \subset P \cap Q$ and the fact that M/L

is torsionfree as a left R/Q-module. As X is left link closed, it follows that $Q \in X$, so M/L is C-torsionfree. It thus follows that K/L is not C-torsion, contradicting the fact that $L \neq 0$.

This proves that $L = K$, indeed. Hence $PK = 0$, i.e., $Ann_R^\ell(K) = P$, as K is C-torsionfree and so $Ann_{R_X}^\ell(K) = R_X P$.

Let us now assume M to be a finitely generated left R_X-module, which is an essential extension of K. Then there exist finitely generated uniform submodules K' resp. M' of K resp. M, such that $K' \subseteq M'$ with $R_X K' = K$ resp. $R_X M' = M$. Clearly $ass_R(M') = ass_R(N') = P$ and N' is torsionfree as a left R/P-module, since $R_X N \hookrightarrow R_X/R_X P$. So, by (6.2), there exist prime ideals $P_1, \ldots, P_n \in X$ with $P_1 \ldots P_n M' = 0$. It follows that $R_X P_1 \ldots R_X P_n M = 0$ and so M is artinian. This shows that R_X is a classical localization.

Let us now verify the converse implication. First, note that the incomparability property follows from the fact that every $R_X/R_X Q$ is artinian and that $Q \subset P$ implies $R_X Q \subset R_X P$.

Next, consider a link $P \rightsquigarrow Q$ in R with $P \in X$. As $C = \mathcal{C}_R(X) \subseteq \mathcal{C}_R(P)$, we also have $C \subseteq \mathcal{C}_R(Q)$. By (6.2), there exists an exact sequence of finitely generated uniform left R-modules

$$0 \to L \to M \to N \to 0,$$

with $L \subseteq R/P$ and $N \subseteq R/Q$. If we localize this exact sequence with respect to C, then we obtain an exact sequence of finitely generated uniform left $C^{-1}R$-modules and $C^{-1}L$ injects into $C^{-1}R/C^{-1}P$, hence it is simple. As R_X is a classical localization, $C^{-1}M$ is artinian, whence so is $C^{-1}N$. As $C^{-1}N$ injects into $C^{-1}R/C^{-1}Q$, it follows that $C^{-1}Q$ is primitive. So $Q \in X$, proving that X is left link closed.

Assume that $P \in X$ does not satisfy the left second layer condition. Then there exists an exact sequence of finitely generated uniform left R-modules

$$0 \to L \to M \to N \to 0$$

such that L is $\mathcal{C}_R(P)$-torsionfree, $Ann_R^\ell(M) = Q \subset P$ and M is $\mathcal{C}_R(Q)$-torsion. Localizing this exact sequence at $C = \mathcal{C}_R(X)$, we obtain an exact sequence of finitely generated left R_X-modules

$$0 \to C^{-1}L \to C^{-1}M \to C^{-1}N \to 0$$

As L is C-torsionfree and essential in M, clearly M is C-torsionfree and uniform as a left R_X-module. Since the localization is classical, $C^{-1}M$ is artinian, so for some $P_1, \ldots, P_n \in X$, the module $C^{-1}M$ is annihilated by $R_X P_1 \ldots P_n$. Hence $P_1 \ldots P_n M = 0$, whence $P_1 \ldots P_n \subseteq Q$. But then $P_i \subseteq Q \subset P$ for some $1 \leq i \leq n$, contradicting the incomparability assumption.

Finally, assume I is a left ideal of R disjoint from $C = \mathcal{C}_R(X)$, but with $I \cap \mathcal{C}_R(P) \neq \varnothing$ for every $P \in X$. We may choose I maximal with respect to this property, in which case $M = R/I$ is not C-torsion, whereas every proper quotient of M is (in particular, M will then be C-torsionfree. This implies that $C^{-1}M$ is simple over $C^{-1}R$, so $Ann_{R_X}^\ell(C^{-1}M) = R_X P$ for some $P \in X$. This implies M to be generated by a $\mathcal{C}_R(P)$-torsion element as an R/P-module, showing that M is $\mathcal{C}_R(P)$-torsion. On the other hand, $C^{-1}M$ injects into $R_X/R_X P$, which contradicts the fact that M is $\mathcal{C}_R(P)$-torsion. This shows that the weak left intersection property holds, which proves the assertion. $\square$

To conclude this Section, let us point out the following remarkable symmetry result, which will be used several times in the sequel:

(6.10) Proposition. [15] *Assume R to be a noetherian prime PI ring. Then $X \subseteq Spec(R)$ is left link closed if and only if X is right link closed. In particular, $X \subseteq Spec(R)$ is (classically) left localizable if and only X is (classically) right localizable.*

7. FBN rings

(7.1) Let us call an ideal I of a ring R *bounded* if and only if every essential left ideal of R/I contains a nonzero ideal. A ring R is *fully left bounded* if and only if every prime ideal of R is left bounded.

(7.2) Lemma. *If R satisfies the descending chain condition on left annihilators, then every left annihilator is of the form $Ann_R^\ell(S)$ for a finite subset S of R.*

Proof. Consider an arbitrary annihilator $Ann_R^\ell(X)$ with $X \subseteq R$. By hypothesis, there is a minimal annihilator of the form $Ann_R^\ell(S)$ for some finite subset S of X. For any other finite subset T of X, it follows that

$$Ann_R^\ell(S \cup T) = Ann_R^\ell(S) \cap Ann_R^\ell(T) \subseteq Ann_R^\ell(S),$$

whence $Ann_R^\ell(S \cup T) = Ann_R^\ell(T)$, by the minimality of $Ann_R^\ell(S)$. We thus find that $Ann_R^\ell(S) = Ann_R^\ell(X)$, indeed. $\qquad\qquad\square$

Of course, the assumption in the previous Lemma is met by any left noetherian ring R. Denote by $Spec(R{-}\mathbf{mod})$ the set of isomorphism classes $[E]$ of nonzero indecomposable injective left R-modules, then we may define a map

$$\Phi : Spec(R{-}\mathbf{mod}) \to Spec(R) : [E] \mapsto ass_R(E),$$

(taking into account the fact that every indecomposable injective left R-module is cotertiary. It is clear that Φ is surjective, as each prime ideal $P \in Spec(R)$ is associated to the indecomposable injective direct summands of the injective hull $E(R/P)$ of R/P.

(7.3) Theorem. [24, 27, 46, 76] *Let R be a left noetherian ring. Then the following assertions are equivalent:*

(7.3.1) *R is fully left bounded;*

(7.3.2) *every cotertiary left R-module is isotypic.*

(7.3.3) *the map* $\Phi : Spec(R-\mathbf{mod}) \to Spec(R)$ *is bijective;*

(7.3.4) *if M is a finitely generated left R-module, then R satisfies the descending chain condition on annihilators of subsets of M;*

(7.3.5) *for every finitely generated left R-module M, there exist elements $x_1, \ldots, x_n \in M$, such that*

$$Ann^{\ell}_R(M) = Ann^{\ell}_R(x_1) \cap \ldots \cap Ann^{\ell}_R(x_n);$$

(this condition is usually referred to as Gabriel's condition (H));

(7.3.6) *every radical in $R-\mathbf{mod}$ is symmetric;*

(7.3.7) *for every prime ideal P of R, the radical σ_P is symmetric (and hence, in particular, $\sigma_P = \sigma_{R\backslash P}$.*

Proof.   To prove that (1) implies (2), consider a cotertiary left R-module M and a decomposition $E(M) = \bigoplus_i E_i$ into indecomposable injectives. Then $ass_R(E_i) = ass_R(M)$ for each E_i, hence the E_i are mutually isomorphic, indeed.

Conversely, to show that (2) implies (1), consider two indecomposable injectives E_1 and E_2 with $ass_R(E_1) = ass_R(E_2) = P$, then $Ass_R(E_1 \oplus E_2) = \{P\}$ as well. So, $E_1 \oplus E_2$ is cotertiary, hence isotypic, hence $E_1 \cong E_2$, indeed.

In order to show that (1) implies (3), let us show that every indecomposable injective left R-module E is uniquely determined by the prime ideal $P = ass(E)$. In order to realize this, it suffices to find a monomorphism $E \hookrightarrow E(R/P)$. Actually, if we decompose $E(R/P)$ into a direct sum of (mutually isomorphic!) indecomposable injectives, say $E(R/P) = \bigoplus_i E_i$, then the Azumaya theorem implies that E is isomorphic to E_i. Now, by definition, we have that $P = Ann^{\ell}_R(x)$ for some $x \in E$ and $L = Ann^{\ell}_R(x)$ is an irreducible left R-ideal containing P. We claim that L/P is not essential within R/P. Indeed, otherwise L/P would contain a twosided ideal I/P with $I \supset P$. But then

$IRx = Ix \subseteq Lx = 0$, contradicting $Ann_R^\ell(Rx) = P$. So L/P is not essential, indeed, hence we may find some left R-ideal $K \supset P$, with the property that $L/P \cap K/P = 0$, i.e., $L \cap K = P$. Writing K as an intersection of irreducible left ideals, we then obtain an irreducible decomposition $P = L \cap L_1 \cap \ldots \cap L_n$. It follows that

$$E_R(R/P) = E_R(R/L) \oplus E_R(R/L_1) \oplus \ldots E_R(R/L_n),$$

and, in particular,

$$E = E_R(Rx) = E_R(R/L) \subseteq E_R(R/P),$$

as desired. Let us now show that (3) implies (4). Let M be a finitely generated left R-module and decompose its injective hull into indecomposable injectives, say $E(M) = E_1 \oplus \ldots \oplus E_n$. For each $1 \le i \le n$, we let $M_i = \pi_i(M)$, where $\pi_i : E(M) \to E_i$ denotes the canonical projection. For any non-empty $X \subseteq M$, we have $Ann_R^\ell(X) = \bigcap_{i=1}^n Ann_R^\ell(\pi_i(X))$. It thus follows that R will satisfy the descending chain condition on annihilators of subsets of M if and only if it satisfies it for annihilators of subsets of each of the M_i. So we may assume M to be finitely generated and uniform, and, under this assumption, M is cotertiary, say $ass_R(M) = P \in Spec(R)$. We claim that if $\varnothing \ne X \subseteq M$, then there exist elements $x_1, \ldots, x_m \in X$, with $Ann_R^\ell(X) = \bigcap_{i=1}^m Ann_R^\ell(x_i)$. Indeed, we may assume without loss of generality that $0 \notin X$. Let $I = Ann_R^\ell(X)$, then for any $Q \in Ass_R(R/I)$ it follows that R/I has a left R-submodule J/I with the property that

$$Q = Ann_R^\ell(J/I) = (I : J) = Ann_R^\ell(JX).$$

Since Q is the annihilator of a submodule of M, it follows that $Q \subseteq P$ and as

$$P = Ann_R^\ell(M) \subseteq Ann_R^\ell(X) = Q \subseteq P,$$

we have $Ass_R(R/I) = \{P\}$. Moreover, for each $x \in X$ we see that $(R/I)(Ann_R^\ell(x)/I) = R/Ann_R^\ell(x) \cong Rx \subseteq M$. So, Rx is uniform and

$ass_R(x) = P$. Finally, since $\bigcap_{x \in X} Ann_R^\ell(x) = I$, (7.2) yields that there exists a finite subset $\{x_1, \ldots, x_m\}$ of X such that $I = Ann_R^\ell(X) = \bigcap_{i=1}^m Ann_R^\ell(x_i)$. Now, let $\{X_\alpha; \alpha \in A\}$ be a family of non-empty subsets of M, with $\bigcup_{\alpha \in A} X_\alpha = X$. Then $Ann_R^\ell(X) = \bigcap_{\alpha \in A} Ann_R^\ell(X_\alpha)$, while on the other hand there exist elements $x_1, \ldots, x_m \in X$ with $Ann_R^\ell(X) = \bigcap_{i=1}^m Ann_R^\ell(x_i)$. Let $B \subseteq A$ be finite and with the property that $\{x_1, \ldots, x_m\} \subseteq \bigcup_{\alpha \in B} X_\alpha$. Then $Ann_R^\ell(X) = \bigcap_{\alpha \in B} Ann_R^\ell(X_\alpha)$. This proves that R satisfies the descending chain condition on left annihilators, indeed. That (4) implies (5) is obvious. Indeed, let M be a finitely generated left R-module and let $I = Ann_R^\ell(M)$. Pick a nonzero element $m_1 \in M$. If $Ann_R^\ell(m_1) = Ann_R^\ell(M)$, we are done. If not, there exists an element $m_2 \in M$, such that

$$Ann_R^\ell(m_1) \supset Ann_R^\ell(m_1) \cap Ann_R^\ell(m_2).$$

If $Ann_R^\ell(m_1) \cap Ann_R^\ell(m_2) = Ann_R^\ell(M)$, we are done. In the other case, we iterate this process. By assumption, the chain thus obtained stops after a finite number of steps and this shows (5) to hold true, indeed.

To prove that (5) implies (6), consider a radical σ in $R-\mathbf{mod}$ and a left ideal $L \in \mathcal{L}(\sigma)$. Then $(L : R) = Ann_R^\ell(R/L)$, hence there exist elements $\overline{a_1}, \ldots, \overline{a_n}$ of R/L such that $(L : R) = \bigcap_{i=1}^n Ann_R^\ell(\overline{a_i})$. As each $Ann_R^\ell(\overline{a_i})$ belongs to $\mathcal{L}(\sigma)$, so does there intersection, hence $(L : R) \in \mathcal{L}(\sigma)$, which shows that σ is symmetric, indeed.

It is clear that (6) implies (7), so let us prove that (7) implies (1) to finish the proof. Consider a prime ideal P of R and a left ideal L such that L/P is essential in R/P. We claim that R/L is $\sigma_{R \setminus P}$-torsion. Indeed, in the contrary case, there would exist a nonzero R-linear homomorphism $f : R/L \to E_R(R/P)$. Since $P \subseteq L$, clearly $f(R/L) \subseteq E_{R/P}(R/P)$. But then $f(1 + L)$ is annihilated by L/P, which contradicts the fact that $E_{R/P}(R/P)$ is a nonsingular left R/P-module, as R/P is a left noetherian prime (and hence Goldie) ring.

By assumption, σ_P is symmetric, hence $(L : R) \in \mathcal{L}(\sigma_P)$. However,

$P \subset (L : R)$, and this inclusion is strict because $P \notin \mathcal{L}(\sigma_P)$. We thus have found a nonzero twosided ideal of R/P contained in L/P, which finishes the proof.

□

The previous result easily shows all commutative noetherian rings to be fully bounded. Somewhat more generally, all left noetherian pi rings are fully left bounded, cf. [72].

It should be clear from their very definition, that FBN rings behave very nicely in many respects. This is also due to the following result, which has already been announced in Chapter I:

(7.4) Lemma. *Every FBN ring satisfies the strong second layer condition.*

Proof. Consider a finitely generated R-module M with $Ann_R^\ell(M) = Q \in Spec(R)$, then we claim that M is prime. Indeed, if not, one may find a left R-submodule $N \subset M$ with $Ann_R^\ell(N) = P = ass_R(M) \supset Q$. Since $0 \neq P/Q$ is essential in R/Q, it contains a regular element, so N is torsion as a left R/Q-module and hence so is M. Since R is an FBN ring,

$$Q = Ann_R^\ell(M) = \bigcap_{i=1}^{n} Ann_R^\ell(m_i),$$

for a finite number of elements, m_i and each of these $Ann_R^\ell(m_i)$ contains a twosided ideal strictly containing Q – a contradiction. □

(7.5) We conclude this Section with a brief study of rings which are module-finite over their (noetherian) center.

First note that if R is finitely generated over its center C and if C is noetherian, then clearly R is noetherian as well. Moreover, it is fairly easy to see that the ring R is then also FBN. Indeed, if M is a finitely generated left R-module, for example, say with generators $m_1, \ldots, m_n$, then we claim that $Ann_R^\ell(M) = \bigcap_{i=1}^{n} Ann_R^\ell(m_i)$. For, take $r \in R$

belongs to this intersection and let $m = \sum_{i=1}^{n} c_i m_i$ for some $c_i \in C$. Then $rm = r \sum_{i=1}^{n} c_i m_i = \sum_{i=1}^{n} c_i r m_i = 0$, hence $r \in Ann_R^\ell(M)$, as claimed.

On the other hand, if R is module-finite over its noetherian center C, then the cliques of R are extremely easy to describe in terms of the prime ideals of C, cf. (7.8) below. The proof of this result may be found in [37]; we will give a brief sketch of it, for the reader's convenience.

(7.6) Lemma. [37, 11.8] *Let R be an indecomposable noetherian ring all of whose prime ideals are maximal. Then $Spec(R)$ consists of a single clique.*

Proof. It is clear that R satisfies the strong second layer condition. Now, if A_1 is a clique and if $A_2 = Spec(R) \setminus A_1$, then it is easy to see that the maximality of all prime ideals implies that $I_1 \cap I_2 = 0$, where, for $i = 1, 2$, the twosided ideal I_i consists of all elements of R annihilated by some product of prime ideals in A_i.

If K/I_1 is a nonzero left submodule of R/I_1, then obviously K/I_1 cannot be annihilated by a prime ideal in A_1, so $Ass_R(R/I) \subseteq A_2$. Since A_2 is link closed, it follows from (6.5) that R/I_1 is annihilated by some product of prime ideals in A_2. So, $R/I_1 + I_2$ is annihilated by a product of prime ideals in A_1 and by symmetry by a product of prime ideals in A_2, i.e., $I_1 + I_2 = R$.

Since R is indecomposable, it follows that $I_1 = R$ or $I_2 = R$, the latter conclusion not occurring (as otherwise some product of prime ideals in A_2 would be zero and $Y = Spec(R)$, which is impossible). So $I_1 = R$, proving that $X = Spec(R)$, indeed.   □

We will also need the following (technical) result:

(7.7) Lemma. *Assume the ring R to be module-finite over its noetherian center C. If I is an ideal of C, then there exists a positive integer*

n, such that every element of R/RI that may be lifted to the center of R/RI^n may also be lifted to C.

Proof. Assume that R is generated by $x_1, \ldots, x_n$ as a C-module and denote by M the finitely generated C-module R^n. Denote by f the C-linear homomorphism

$$f : R \to M : r \mapsto (x_1 r - r x_1, \ldots, x_n r - r x_n).$$

Since I has the Artin-Rees property as C is noetherian, there exists some positive integer m with the property that $f(R) \cap I^m M \subseteq I f(R) = f(RI)$. A class $r + RI \in R/RI$ that lifts to the center of R/RI^m has the property that $x_i r - r x_i \in RI^m$ for any $1 \leq i \leq n$, hence $f(r) \in I^m M$. As $f(R) \cap I^m M \subseteq f(RI)$, we find that $f(r) = f(x)$ for some $x \in RI$. So, $r - x \in Ker(f) = C$, and since $r - x + RI = r + RI$, the shows that we have lifted $r + RI$ to an element of C, indeed. $\square$

We may now finally prove our main result:

(7.8) Theorem. *([37, 11.7], [64]) Let R be module-finite over its noetherian center C and let P be a prime ideal of R. Then*

$$Cl(P) = \{Q \in Spec(R);\ Q \cap C = P \cap C\}.$$

Proof. One inclusion is clear. Indeed, if $c \in C$ belongs to P, then $c \in Q$ for any $Q \in Spec(R)$ with $Q \rightsquigarrow P$ or $P \rightsquigarrow Q$ (hence for any $Q \in Cl(P)$). For, if we assume $Q \rightsquigarrow P$, defined through the twosided ideal I, for example, then $(Q \cap P)P \subseteq I$, as $Ann_R^r(Q \cap P/I) = P$, by definition, hence $c(Q \cap P) = (Q \cap P)c \subseteq I$. It follows that $c \in Ann_R^l(Q \cap P/I) = Q$, which proves the assertion.

To prove the other inclusion, let $\mathfrak{p} = P \cap C$ and $S = C \setminus \mathfrak{p}$, then it is clear that $S^{-1}C$ is the center of $S^{-1}R$. From (1.9) in Chapter I and the first part of the proof, it follows that each of the extended ideals $Q^e = C^{-1}Q$ with $Q \cap C = \mathfrak{p}$ belongs to $Spec(R)$ and it is clear that $Q^e \cap S^{-1}C = P^e \cap S^{-1}C = S^{-1}\mathfrak{p}$. Using this and the (easy to verify) fact that $Q^e \rightsquigarrow P^e$ if and only if $Q \rightsquigarrow P$, one may thus reduce the proof to that $P \cap C$ is a maximal ideal $\mathfrak{m}$ of C.

From the previous Lemma, it follows that there exists some positive integer n with the property that any element of $R/R\mathfrak{m}$ that lifts to the center of $R/R\mathfrak{m}^n$ also lifts to C. From this, one may deduce that the ring $R/R\mathfrak{m}^n$ is indecomposable. Indeed, if $e = a + R\mathfrak{m}^n$ is a central idempotent in $R/R\mathfrak{m}^n$, then $a + R\mathfrak{m}$ thus lifts to C, i.e., $a + R\mathfrak{m} = b + R\mathfrak{m}$ for some $b \in C$, with $b - b^2 \in R\mathfrak{m} \cap C = \mathfrak{m}$. As $C/\mathfrak{m}$ is a field, clearly b or $1 - b$ belongs to $\mathfrak{m}$, so a or $1 - a$ belongs to $R\mathfrak{m}$, implying that e or $1 - e$ belongs to $R\mathfrak{m}/R\mathfrak{m}^n$. But then e or $1 - e$ is nilpotent, i.e., $e = 0$ or $e = 1$.

To conclude, note that $R/R\mathfrak{m}^n$ is artinian (being module-finite over the artinian ring $C/\mathfrak{m}^n$) and apply (7.6) to infer that all prime ideals of $R/R\mathfrak{m}^n$ are in the same clique. Since one easily deduces from this that all prime ideals of R containing $R\mathfrak{m}^n$ belong to the same clique (see also (7.18) in the next Chapter), this proves the assertion.   □

Chapter II

EXTENSIONS

One of the many techniques used in commutative algebra to study the structure of a commutative ring R consists in considering well-chosen over-rings S of R (local rings obtained by localizing at a prime ideal P of R, or the integral closure of R in some extension field, for example) and try to infer information on R from properties of the special ring S. From a geometric point of view, one studies the affine scheme $Spec(R)$ through properties of well-chosen affine schemes $Spec(S)$ and morphisms $Spec(S) \to Spec(R)$.

In the noncommutative case, similar techniques have been developed, cf. [4, 87, et al], but the geometric study of R through its prime spectrum $Spec(R)$ does not work so elegantly, in particular, because a ring extension $R \hookrightarrow S$ does not necessarily induce a map $Spec(S) \to Spec(R)$.

Indeed, to give an example, consider the extension

$$R = K \times K \cong \begin{pmatrix} K & 0 \\ 0 & K \end{pmatrix} \hookrightarrow \begin{pmatrix} K & K \\ K & K \end{pmatrix} = S,$$

for some field K. Clearly, the zero-ideal $M_2(0)$ is prime in $S = M_2(K)$, whereas $\{0\} \times \{0\}$, its inverse image in R, is not prime anymore. So, here (unlike what happens in the commutative case), an inclusion $R \subseteq S$

does not induce a map $Spec(S) \to Spec(R)$, by taking intersections with R.

In order to still be able to use similar techniques as in the commutative case, one is thus naturally invited to put some restrictions on the inclusions $R \subseteq S$ or the more general ring morphisms $R \to S$.

In this Chapter, we will study several quite general types of ring extensions, each one allowing us to get some grasp on the behaviour of prime ideals. In particular, besides topological implications, we will also be interested in the interplay between these extensions and general localization theory with respect to radicals or, more concretely, at prime ideals.

1. Centralizing extensions

(1.1) Let us start with some definitions. An R-bimodule M (i.e., a twosided R-module with compatible left and right R-actions) is said to be *centralizing* (or sometimes an *R-bimodule in the sense of Artin* [3]), if M is generated as a left (or, equivalently, as a right) R-module by M^R, the set of all $m \in M$ with the property that $rm = mr$ for all $r \in R$. Clearly, this is equivalent to requiring the existence of an index set I and a surjective morphism $\pi : R^{(I)} \to M$ in $R-\mathbf{mod}-R$, the category of R-bimodules. Indeed, if M is a centralizing R-bimodule, then the surjective map

$$R^{(M^R)} \to M : (r_m; \ m \in M^R) \mapsto \sum_{m \in M^R} r_m m = \sum_{m \in M^R} m r_m$$

works as indicated. Conversely, in the presence of a surjective R-bimodule map $\pi : R^{(I)} \to M$, the R-bimodule M is generated on the left (and on the right) by $\{\pi(e_i); \ i \in I\} \subset M^R$, where $\{e_i; \ i \in I\}$ is the canonical basis of $R^{(I)}$.

(1.2) Although the category of centralizing R-bimodules is obviously not abelian, in general (due to the lack of existence of kernels, for example), in [87], the authors developed a complete localization theory within the category of these. One of the applications of this set-up consists in the construction of reasonably well-behaving structure sheaves, at least over the prime spectrum of rings satisfying a polynomial identity. We will not go into this here, but instead refer to [87] and the literature mentioned there.

The reason why centralizing R-bimodules (and centralizing extensions, as we will see in a moment) may be used in the present context essentially stems from the following, elementary result:

(1.3) Lemma. *If M is a centralizing R-bimodule, then we have $IM = MI$, for any twosided ideal I of R.*

Proof. Indeed, any $m \in M$ may be written as $m = \sum r_\alpha m_\alpha = \sum m_\alpha r_\alpha$ for certain $r_\alpha \in R$ and $m_\alpha \in M^R$. If $i \in I$, then we thus obtain

$$im = i \sum r_\alpha m_\alpha = \sum (ir_\alpha)m_\alpha = \sum m_\alpha (ir_\alpha) \in MI$$

i.e., $IM \subseteq MI$. By symmetry, this yields $IM = MI$, as claimed. $\square$

(1.4) A ring morphism $\varphi : R \to S$ is said to be *centralizing* or a *centralizing extension*, if S is a centralizing R-bimodule for the structure induced by φ. So, φ will be a centralizing extension, if $S = \varphi(R)S^R = S^R\varphi(R)$, where S^R consists of all $s \in S$ with the property that $s\varphi(r) = \varphi(r)s$ for all $r \in R$.

It is clear that for any ring R, the canonical inclusion $R \hookrightarrow R[X]$ (where X is a central variable over R) yields a centralizing extension. More generally, any ring morphism $\varphi : R \to S$ with $S = \varphi(R)Z(S) = Z(S)\varphi(R)$ (where $Z(S)$ is the center of S), is a centralizing extension. In particular, it thus follows that any surjective morphism $R \to S$ is centralizing.

It is also clear that any inclusion $R \subset R\{X_\alpha; \, \alpha \in A\}$, where the X_α

do not mutually commute, but commute with R, yields a centralizing extension (which is not central if $|A| > 1$, however).

Of course, it is trivial to see that any centralizing extension $\varphi : R \hookrightarrow S$ is of the form

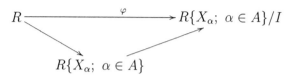

for some index set A and some twosided ideal I of $R\{X_\alpha; \ \alpha \in A\}$. Indeed, we may take $A = S^R$, choosing for I the kernel of the canonical surjection

$$R\{X_s; \ s \in S^R\} \to S : X_s \mapsto s.$$

(1.5) Centralizing extensions share many nice properties with respect to the behaviour of twosided ideals and, in particular, of prime ideals:

(1.5.1) *If $\varphi : R \to S$ is a centralizing extension, then for any twosided ideal I of R we have $S\varphi(I) = \varphi(I)S$. In particular, it follows that $S\varphi(I)$ is a twosided ideal of S.*

Indeed, this follows immediately from (1.3).

(1.5.2) *If $\varphi : R \to S$ is a centralizing extension, then for any $Q \in Spec(S)$ we have $P = \varphi^{-1}(Q) \in Spec(R)$.*

For, if I and J are twosided ideals of R with the property that $IJ \subseteq Q$, then by definition $\varphi(I)\varphi(J) = \varphi(IJ) \subseteq Q$. It follows that

$$\varphi(I)S\varphi(J) = S\varphi(I)\varphi(J) \subseteq Q,$$

so $\varphi(I) \subseteq Q$ or $\varphi(J) \subseteq Q$, as $Q \in Spec(S)$. Hence $I \subseteq P$ or $J \subseteq P$ proving that $P = \varphi^{-1}(Q)$ is prime, indeed.

(1.6) For any twosided ideal I of R, the subsets $\mathcal{K}(\sigma_I)$ resp. $\mathcal{Z}(\sigma_I)$ of $Spec(R)$ defined in (4.8) in Chapter I, will be denoted by $X_R(I)$ resp. $V_R(I)$ and are given by

$$X_R(I) = \{P \in Spec(R); \ I \not\subseteq P\}$$

resp.

$$V_R(I) = \{P \in Spec(R);\ I \subseteq P\}.$$

Let us write $X(I) = X_R(I)$ resp. $V(I) = V_R(I)$, if no ambiguity arises.

It is then clear that $X(I)$ only depends upon (I), the twosided ideal generated by I (or even $rad(I)$, the radical of I!), so we may assume without loss of generality, that I is a twosided ideal of R. It is clear that $X(R) = Spec(R)$ and $X(0) = \varnothing$. Since we also have

$$X(IJ) = X(I) \cap X(J)$$

resp.

$$X(\sum_{\alpha \in A} I_\alpha) = \bigcup_{\alpha \in A} X(I_\alpha)$$

for any pair of twosided ideals I, J resp. any family of twosided R-ideals $\{I_\alpha;\ \alpha \in A\}$, it follows that the family of all $X(I)$, where I is a twosided ideal of R, forms a topology on $Spec(R)$, the so-called *Zariski topology*. The sets of the form $V(I)$ are the closed subsets for this topology.

The following easy result is sometimes very useful if one wants to calculate prime ideals:

(1.7) Lemma. *Let R be a subring of S and assume T to be a common twosided ideal of R and S. Then:*

(1.7.1) *for every nonzero prime ideal P of R, there exists a prime ideal P' of S such that $P' \cap R \subseteq P$;*

(1.7.2) *if $T \not\subseteq P$, then there exists a unique prime ideal P' of S such that $P' \cap R = P$.*

Proof. Let P' be a twosided ideal of S, which is maximal with the property that $P' \cap R \subseteq P$. We claim that $P' \in Spec(S)$. Indeed, assume that $x, y \in S \setminus P'$, while $xSy \subseteq P'$. Then, by the maximality of P', we have $(P' + Rx) \cap R \not\subseteq P$ and $(P' + Ry) \cap R \not\subseteq P$. However,

$$((P'+Rx)\cap R)((P'+Ry)\cap R) \subseteq (P'+Rx)(P'+Ry)\cap R \subseteq P'\cap R \subseteq P,$$

a contradiction. This shows that P' is a prime ideal of S.

On the other hand, let us now assume that $T \not\subseteq P$, and consider the twosided ideal TPT of S. Clearly, $TPT \cap R \subseteq P$, so by (1), there exists some $P' \in Spec(S)$ with $TPT \subseteq P'$. But then $TSPST \subseteq P'$, hence $SPS \subseteq P'$, as $T \not\subseteq P'$. So, $P \subseteq P' \cap R$, which implies that $P = P' \cap R$, indeed.

Finally, if P'' is another prime ideal of S with $P = P'' \cap R$, then $TP'' \subseteq P'' \cap R = P \subseteq P'$, so $P'' \cap P'$. By symmetry, we also have $P' \subseteq P''$, whence equality.   □

Note that we have actually proved that any prime ideal P' with $T \not\subseteq P'$ intersects R in a prime ideal $P' \cap R$.

(1.8) As an application, choose a commutative domain C, with subrings A and B and ideals I resp. J with $IJ \subseteq A \cap B$ and consider the ring

$$R = \begin{pmatrix} A & I \\ J & B \end{pmatrix}.$$

If both I and J are nonzero, then R is a domain and if C is finite over A and B, then R is noetherian. In order to calculate the prime ideals of R, one applies the previous result with $S = M_2(C)$ and with $T = M_2(IJ)$ as a common twosided ideal. The prime ideals of R which do not contain T are thus of the form $[P] = P \cap R$ for some $P \in Spec(M_2(R))$ with $T \not\subseteq P$, and these are in bijective correspondence with the prime ideals of C not containing IJ. Put

$$I_1 = \begin{pmatrix} A \cap I & I \\ IJ & B \cap I \end{pmatrix} \text{ resp. } J_1 = \begin{pmatrix} A \cap J & IJ \\ J & B \cap J \end{pmatrix}.$$

Clearly $I_1 J_1 \subseteq M_2(IJ)$. So, if a prime ideal of R contains T, it must contain either I_1 or J_1. From this, it thus easily follows that the prime ideals of R are thus of the form:

- $[P] = M_2(\mathfrak{p}) \cap R$, with $\mathfrak{p} \in X_C(IJ)$;

- $\mathfrak{q}_+ = \begin{pmatrix} \mathfrak{q} & I \\ J & B \end{pmatrix}$, with $\mathfrak{q} \in V_A((I \cap A)(J \cap A))$;

- $\mathfrak{q}_- = \begin{pmatrix} A & I \\ J & \mathfrak{q} \end{pmatrix}$, with $\mathfrak{q} \in V_B((I \cap B)(J \cap B))$.

(1.9) As an example, let us start from the polynomial ring $D = \mathbf{C}[X]$ and the maximal ideal $\mathfrak{m} = (X)$ of D. Consider the ring

$$R = \begin{pmatrix} D & \mathfrak{m} \\ \mathfrak{m} & D \end{pmatrix}$$

From the previous discussion, it follows that $Spec(R)$ consists of the zero-ideal (0), the maximal ideals $M_\alpha = R(X - \alpha)$ and the two *exceptional* maximal ideals

$$M_+ = \begin{pmatrix} \mathfrak{m} & \mathfrak{m} \\ \mathfrak{m} & D \end{pmatrix} \text{ resp. } M_- = \begin{pmatrix} D & \mathfrak{m} \\ \mathfrak{m} & \mathfrak{m} \end{pmatrix}.$$

Clearly, the center $Z(R)$ of R is just the ring D, embedded diagonally. It is obvious that for any $0 \neq \alpha \in \mathbf{C}$, we have $M_\alpha \cap D = (X - \alpha)$, whereas $M_+ \cap D = M_- \cap D = (X)$. The prime spectrum $Spec(R)$ may thus graphically be represented as

All points in $Spec(R)$ different from the zero-ideal are closed in the Zariski topology, as they correspond to maximal ideals, the zero-ideal itself being dense in $Spec(R)$.

One of the reasons why we are so interested in centralizing morphisms is given by the next result:

(1.10) Proposition. *Any centralizing extension* $\varphi : R \to S$ *induces a continuous map*

$$^{a}\varphi : Spec(S) \to Spec(R) : Q \mapsto \varphi^{-1}(Q),$$

with respect to the Zariski topology.

Proof. We have already verified that φ induces a map $^{a}\varphi : Spec(S) \to Spec(R)$, indeed. To finish the proof, we will show that for any twosided ideal I of R we have

$$(^{a}\varphi)^{-1}(X_R(I)) = X_S(\varphi(I)) = X_S(S\varphi(I)).$$

Now, clearly $Q \in (^{a}\varphi)^{-1}(X_R(I))$ if and only if $\varphi^{-1}(Q) = (^{a}\varphi)(Q) \in X_R(I)$, i.e., $I \not\subset \varphi^{-1}(Q)$. Since this is equivalent to $\varphi(I) \not\subset Q$, i.e., $Q \in X_S(\varphi(I))$, this proves the assertion. $\qquad\square$

(1.11) Example. Consider the diagonal embedding

$$j : D \hookrightarrow R = \begin{pmatrix} D & \mathfrak{m} \\ \mathfrak{m} & D \end{pmatrix},$$

where $D = \mathbf{C}[X]$ and $\mathfrak{m} = (X) \le D$. With notations as before, for any $0 \ne \alpha \in \mathbf{C}$, we have

$$(^{a}j)(M_{\alpha}) = j^{-1}(M_{\alpha}) = M_{\alpha} \cap D = (X - \alpha),$$

whereas

$$(^{a}j)(M_{+}) = (^{a}j)(M_{-}) = \mathfrak{m}.$$

The morphism $^{a}j : Spec(R) \to Spec(D)$ induced by j may thus graphically be represented as:

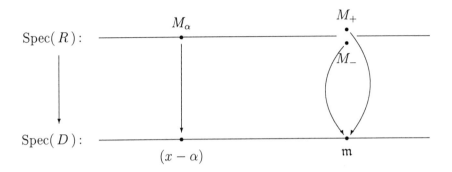

(1.12) As a special case, let us mention that if I is a twosided ideal of R, then the morphism $Spec(R/I) \to Spec(R)$ induced by the canonical surjection $R \to R/I$ may be identified with the canonical inclusion $V_R(I) \subseteq Spec(R)$. To see this, it suffices to verify that mapping any $P \in V(I)$ (i.e., with $I \subseteq P$), to $P/I \in Spec(R/I)$ yields a homeomorphism $V(I) \cong Spec(R/I)$.

We leave the proof of these remarks as a straightforward exercise to the reader.

2. Normalizing extensions

(2.1) Let us generalize the notions introduced in the previous Section as follows. We call an R-bimodule *normalizing*, if it is generated as a left or, equivalently, as a right R-module, by $N_R(M)$, the set of all $m \in M$ with the property that $Rm = mR$. If $M = R$, then we will just write $N(R)$ instead of $N_R(R)$. A ring morphism $\varphi : R \to S$ is said to be *normalizing* or to be a *normalizing extension*, if φ makes S into a normalizing R-bimodule. In other words, this means that $S = \varphi(R)N_R(S) = N_R(S)\varphi(R)$, where $N_R(S)$ consists of all $s \in S$ with the property that $\varphi(R)s = s\varphi(R)$. If a finite number of elements in $N_R(S)$ are sufficient to generate S, then we speak of a *finite* normalizing extension.

Clearly, any centralizing R-bimodule resp. any centralizing extension is a normalizing R-bimodule resp. a normalizing extension.

(2.2) As a typical example of normalizing extensions, let us mention so-called automorphic extensions. Recall that an element m of an R-bimodule M is said to be *R-automorphic*, if there exists an automorphism φ of R with the property that $rm = m\varphi(r)$ for all $r \in R$. We then call an injective ring morphism $R \subseteq S$ *automorphic* if S is generated as an R-module by R-automorphic elements. Somewhat more generally, a not necessarily injective ring extension $\varphi : R \to S$ is said to be *automorphic* if the inclusion $\varphi(R) \subseteq S$ is automorphic.

It is clear that any automorphic extension is normalizing and that any centralizing extension is automorphic (where φ is the identity map, of course), all of the implications being strict, in general.

(2.3) The study of (finite) normalizing extension has been the object of increasing research during the last decade. In order to prove (2.4) below – a key result throughout this text, which contains most of the fundamental properties concerning normalizing extensions – we will need some preliminaries first. Since the result is easily accessible in the literature, we will restrict to a brief sketch of the essentials.

Throughout, let us assume S to be a finite normalizing extension of R, say of the form $S = \sum_{i=1}^{n} Ra_i$, with $a_i \in N_R(S)$ for $1 \le i \le n$. Our starting point will be the observation that any essential left R-submodule N of $M \in S-\mathbf{mod}$ contains an essential left S-submodule. Actually, $L = \bigcap_{i=1}^{n}(N : a_i)$ is easily checked to do the trick.

Let us now verify that for any $M \in S-\mathbf{mod}$, we have that M is noetherian as a left S-module if and only if it is as a left R-module. Indeed, if M is noetherian in $S-\mathbf{mod}$ and not in $R-\mathbf{mod}$, then we may pick a maximal left S-module $L \subseteq M$ such that M/L is not left noetherian in $R-\mathbf{mod}$. So, we may assume that M is not noetherian in $R-\mathbf{mod}$, but that M/L is, for any proper left S-submodule $L \subset M$. Choose a maximal non–finitely–generated left R-submodule $N \subseteq M$. Then N is an essential left R-submodule of M, hence the previous remarks yield an essential left S-submodule $L \subset M$ within N. Now M/L is finitely generated in $R-\mathbf{mod}$ and as L is finitely generated as a left S-module, N is finitely generated as well – a contradiction. Since the other implication is obvious, this proves the assertion.

Next, if $M \in S-\mathbf{mod}$ is simple, then M does not contain any essential left R-submodule by the remarks above, so M is semisimple as a left R-module. On the other hand, if M is left noetherian over S, then we have showed that it is also left noetherian over R and its length may be bound, essentially because $(N : a_i)$ is either maximal or improper for any maximal left R-submodule N of M. Finally, if S is left artinian, then it is of finite length, hence so is R, and it follows that R is left artinian as well.

We may now formulate:

(2.4) Theorem. [12, 13, 37, 41, 42, 50, 52, et al] Let $R \subseteq S$ be a finite normalizing extension, say $S = \sum_{i=1}^{n} Ra_i$ for some positive integer n and some $a_i \in N_R(S)$. Then

(2.4.1) *the ring R is (left) noetherian resp. artinian if and only if S is;*

(2.4.2) *for any prime ideal Q of S, the ideal $Q \cap R$ is semiprime in R and is the intersection of at most n prime ideals $P_1, \ldots, P_n$ minimal over $Q \cap R$; (we will then say that Q lies over each of the P_i);*

(2.4.3) *(Lying over) for each prime ideal P of R, there exists at least one and at most n prime ideals of S lying over P;*

(2.4.4) *(Going up) for any prime ideal Q of S, and any prime ideal P' of R with $Q \cap R \subseteq P'$, there exists a prime ideal Q' of S lying over P' and with $Q \subseteq Q'$;*

(2.4.5) *(Incomparability) for any pair of prime ideals $Q_1 \subseteq Q_2$ of S with $Q_1 \cap R = Q_2 \cap R$, we have $Q_1 = Q_2$.*

Proof. The first statement has been proved by the preceding remarks. To prove the other statements, first note that for any R-subbimodule $B \subseteq S$, the *bound* of B (i.e., the largest twosided ideal of S contained in B) may be given by

$$b(B) = \bigcap \{s \in S;\ \forall 1 \leq i, j \leq n,\ a_i S a_j \in B\} = \bigcap_{i=1}^{n} (B : a_i).$$

Let $Q \in Spec(R)$, then we claim that $Q \cap R$ is semiprime and the intersection of at most n minimal prime ideals $P_1, \ldots, P_n$ over $Q \cap R$. Indeed, one easily reduces to the case $Q = 0$. Choose B to be an R-S-subbimodule of S, which is maximal with respect to $b(B) = 0$. Then, for any $1 \leq i \leq n$, we have $(B : a_i) = S$ if and only if $a_i \in B$ and $(B : a_i)$ is maximal in the lattice of R-S-subbimodules of S with bound zero if $a_i \notin B$. We have $\bigcap_{i=1}^{n}(B : a_i) = b(B) = 0$ and as one easily checks $(B : a_i) \cap R = P_i$ to be prime in R, it follows that

$$\bigcap_{i=1}^{n} P_i = \bigcap_{i=1}^{n}((B : a_i) \cap R) \subseteq \bigcap_{i=1}^{n}(B : a_i) = b(B) = 0,$$

which proves (2).

It is fairly obvious that $b(B)$ is an essential R-subbimodule of S, whenever B is. Using this, one now easily checks the *Lying over* property. Indeed, let $P \in Spec(R)$ and choose an ideal I of S, maximal with respect to $I \cap R \subseteq P$. It is easy to see that I is prime, so it remains to show that P is maximal over $I \cap R$. We may again pass to the case $I = 0$. If P is not minimal, then we may find $P' \in Spec(R)$ with $P' \subset P$. Now P is easily checked to be essential in R. Let B be an R-subbimodule of S, maximal with the property $B \cap R = 0$. Then $B \oplus P$ is an essential R-subbimodule of S, hence so is $b(B \oplus P)$. So, $0 \neq b(B \oplus P) \cap R \subseteq (B \oplus P) \cap R = P$ – a contradiction.

Finally, *Going up* and *Incomparability* are proved in the "usual" way. Indeed, to prove (4), consider $Q \in Spec(R)$ and pick $P' \in Spec(R)$ with $Q \cap R \subseteq P'$. Passing over to the quotients $R/Q \cap R \hookrightarrow S/Q$, we may assume that $Q = 0$ and by *Lying over*, there exists some $Q' \in Spec(S)$ lying over P', which proves *Going up*.

Next, let $Q_1 \subseteq Q_2$ be prime ideals of S with $Q_1 \cap R = Q_2 \cap R$. We may again assume $Q_1 = 0$, i.e., S to be prime. If $Q = Q_2 \neq 0$, then there exists some S-regular element $c \in Q$ and from $Rc \subseteq Q$, one easily deduces Q to be an essential left R-submodule of S, so $Q \cap R \neq 0$ – a contradiction, which proves *Incomparability*. □

We leave it as easy exercise to the reader to give a more precise version of these results, when $R \subseteq S$ is centralizing (in this case, for any ideal Q of S, the intersection $Q \cap R$ is prime in R and $Q \in Spec(S)$ lies over $P \in Spec(R)$, exactly when $Q \cap R = P$). Note also that the second statement in the previous result remains valid if we no longer assume the normalizing extension $R \subseteq S$ to be finite (without the limit about the maximal number of primes minimally lying over $Q \cap R$, of course), cf. [37].

(2.5) The previous result may be given a nice geometric interpretation. Indeed, consider a finite normalizing extension $\varphi : R \to S$, where we assume R (and hence S) to be left noetherian, for simplicity's sake.

If Q is a prime ideal of S, then its inverse image $\varphi^{-1}(Q)$ in R is not necessarily prime, but it is at least semiprime, as we have just seen. It follows that $\varphi^{-1}(Q)$ may be written as an essentially unique irredundant intersection $\varphi^{-1}(Q) = P_1 \cap \ldots \cap P_n$ of prime ideals of R. We may thus define a multivalued map or *correspondence*

$$^a\varphi : Spec(S) \multimap Spec(R),$$

by mapping $Q \in Spec(S)$ to the (finite) set $\{P_1, \ldots, P_n\} \subseteq Spec(R)$. Of course, if φ is a centralizing extension, for example, then $^a\varphi$ is just the usual induced continuous map between the corresponding prime spectra.

Let us call a correspondence $f : X \multimap Y$ between arbitrary topological spaces *continuous* if it has the property that for any open subset $V \subseteq Y$ the inverse image

$$f^{-1}(V) = \{x \in X;\ f(x) \cap V \neq \varnothing\}$$

is open in X. The reader should be well aware of the fact that this does *not* imply the inverse image of a closed subset of Y to be closed in X – an easy example will be given below.

We then have:

(2.6) Proposition. *Any finite normalizing extension* $\varphi : R \to S$ *induces a continuous correspondence*

$$^a\varphi : Spec(S) \multimap Spec(R).$$

Proof. The definition of $^a\varphi$ has already been given above, so let us verify $^a\varphi$ to be continuous. In fact, let us show that if I is a twosided ideal of R, then $(^a\varphi)^{-1}(X_R(I)) = X_S(J)$, where J is the twosided ideal of S generated by the image of I.

If $Q \in (^a\varphi)^{-1}(X_R(I))$, then, by definition, there exists some $P \in$

$Spec(R)$ with $I \not\subset P$ lying over $\varphi^{-1}(Q)$. We then clearly obtain that $J \not\subset Q$, so $Q \in X_S(J)$.

Conversely, if $Q \in X_S(J)$, i.e., if $J \not\subset Q$, then $I \not\subset P$. It then follows that $Q \in ({}^a\varphi)^{-1}(X_R(I))$, which proves the assertion. $\qquad\qquad\square$

(2.7) Let us now again work in the injective case, i.e., consider a finite normalizing extension $i : R \hookrightarrow S$. The *Lying over* property then amounts to the surjectivity of the induced correspondence ${}^ai :$ $Spec(S) \multimap Spec(R)$; in other words, for any $P \in Spec(R)$, there exists some $Q \in Spec(R)$ with $P \in ({}^a\varphi)(Q)$.

Similarly, the *Going up* property yields that ${}^ai$ is closed. Indeed, we claim that $({}^ai)(V_S(J)) = V_R(J \cap R)$ for any twosided ideal J of S.

Let $P \in ({}^ai)(V_S(J))$, then there exists some $Q \in V_S(J)$, i.e., with $J \subseteq Q$, such that P minimally contains $Q \cap R$. So, $J \cap R \subseteq P$, i.e., $P \in V_R(J \cap R)$.

Conversely, let $P \in V_R(J \cap R)$, i.e., $J \cap R \subseteq P$. We may assume without loss of generality J to be semiprime, say $J = Q_1 \cap \ldots \cap Q_m$, for some prime ideals $Q_i \in Spec(S)$. Since

$$(Q_1 \cap R) \ldots (Q_m \cap R) \subseteq (Q_1 \cap R) \cap \ldots \cap (Q_m \cap R) = J \cap R \subseteq P,$$

it follows that $Q_1 \cap R \subseteq P$, for example. By *Going up*, there exists some $Q \in Spec(S)$, lying over P and with $Q_1 \subseteq Q$. But then $Q \in V_S(J)$ and $P \in ({}^ai)(Q)$, so $P \in ({}^ai)(V_S(J))$, indeed.

(2.8) Example. Consider the canonical inclusion

$$i : \begin{pmatrix} D & \mathfrak{m} \\ \mathfrak{m} & D \end{pmatrix} = R \hookrightarrow S = \begin{pmatrix} D & D \\ D & D \end{pmatrix},$$

where $D = \mathbb{C}[X]$ and $\mathfrak{m} = (X) \leq D$. Since S is generated as an R-module by the normalizing elements

$$\begin{pmatrix} 1 & 0 \\ 0 & 1 \end{pmatrix} \text{ and } \begin{pmatrix} 0 & 1 \\ 1 & 0 \end{pmatrix},$$

the extension $i : R \hookrightarrow S$ is normalizing.

Since S is a full matrix ring over D, its prime ideals are in one-to-one correspondence with those of D. It thus follows that $Spec(S)$ consists of the zero-ideal and the maximal ideals $N_\alpha = (X - \alpha)S$, where $\alpha \in \mathbf{C}$. The associated correspondence $^a i : Spec(S) \multimap Spec(R)$ thus maps $N_\alpha \in Spec(S)$ to $M_\alpha \in Spec(R)$ for $0 \neq \alpha \in \mathbf{C}$ and the maximal ideal N_0 to the pair $\{M_+, M_-\}$ (with notations as in (1.9)).

In a picture:

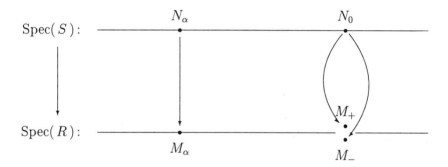

(2.9) Example. As another typical example, consider the canonical inclusion

$$j : \begin{pmatrix} D & 0 \\ 0 & D \end{pmatrix} = R \hookrightarrow S = \begin{pmatrix} D & \mathfrak{m} \\ \mathfrak{m} & D \end{pmatrix},$$

with notations as before. The morphism j is a finite normalizing extension, as S is generated as an R-module by the normalizing elements

$$\begin{pmatrix} 1 & 0 \\ 0 & 1 \end{pmatrix} \text{ and } \begin{pmatrix} 0 & X \\ X & 0 \end{pmatrix}.$$

The ring R may be identified with $D \times D$, so $Spec(R) = Spec(D) \sqcup Spec(D)$. In particular, $Spec(R)$ consists of the prime ideals

$$(X - \alpha)_+ = \begin{pmatrix} (X - \alpha) & 0 \\ 0 & D \end{pmatrix} \text{ resp. } (X - \alpha)_- = \begin{pmatrix} D & 0 \\ 0 & (X - \alpha) \end{pmatrix},$$

where $\alpha \in \mathbf{C}$. With this identification, for any $0 \neq \alpha \in \mathbf{C}$, the associated correspondence ${}^{a}j : Spec(S) \multimap Spec(R)$ maps $M_\alpha \in Spec(S)$ to the pair $\{(X - \alpha)_+, (X - \alpha)_-\}$, and M_+ resp. M_- to $(X)_+$ resp. $(X)_-$.

In a picture:

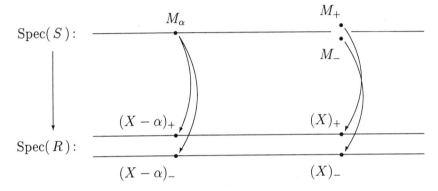

Note that the set $V_R((X)_+) \subseteq Spec(R)$ is closed, whereas

$$({}^{a}j)^{-1}(V_R((X)_+)) = Spec(S) \setminus \{M_+\}$$

is open (and *not* closed!).

3. Strongly normalizing extensions

(3.1)   Although, as we pointed out in the previous Section, normalizing extensions $\varphi : R \to S$ have been used with success in many applications within ring theory and representation theory, their main drawback in geometric applications is that (even in the finite case) they at best lead to *multivalued* maps

$$Spec(S) \dashrightarrow Spec(R) : Q \mapsto \{P_1, \ldots, P_n\},$$

where $P_1, \ldots, P_n$ are the prime ideals of R minimally lying over the semiprime ideal $\varphi^{-1}(Q)$.

On the other hand, (geometrically well-behaving) centralizing extensions are usually not general enough to be applied to purely ring theoretical problems. This motivates us to introduce and study strongly normalizing bimodules and extensions.

(3.2)   Let us define an R-bimodule M to be *strongly normalizing*, if $M = RN_R^s(M) = N_R^s(M)R$, where $N_R^s(M)$ is the set of all $m \in M$ with the property that $Im = mI$ for any twosided ideal I of R. A ring morphism $\varphi : R \to S$ is said to be *strongly normalizing* or to be a *strongly normalizing extension* if φ makes S into a strongly normalizing R-bimodule, i.e., if $S = \varphi(R)N_R^s(S) = N_R^s(S)\varphi(R)$, where $N_R^s(S)$ consists of all $s \in S$ with the property that $\varphi(I)s = s\varphi(I)$ for any twosided ideal of R. It then easily follows that $\varphi(N(R)) \subseteq N(S)$, but, in contrast with centralizing extensions, we do not necessarily have $\varphi(Z(R)) \subseteq Z(S)$.

We will see below that the notion of a strongly normalizing bimodule resp. extension is exactly the notion we want in order to be able to work efficiently algebraically as well as geometrically.

Let us already point out that we have implications

$$(\text{centralizing}) \Rightarrow (\text{strongly normalizing}) \Rightarrow (\text{normalizing})$$

It is easy to see that each of these implications is strict.

The following Lemma gives a typical example of a strongly normalizing extension (which is not centralizing, in general). Recall that if R is a ring and if α is an automorphism of R, then the *skew polynomial ring* $R[X, \alpha]$ is defined by putting $Xr = \alpha(r)X$. We then have:

(3.3) Lemma. *If R is a ring and if α is the inner automorphism of R associated to some invertible element $a \in R$, then the canonical inclusion $R \hookrightarrow R[X, \alpha]$ is strongly normalizing.*

Proof. Obviously $R[X, \alpha]$ is generated as a left (and right) R-module by the family $\{X^n; \ n \in \mathbf{N}\}$ and we claim these elements to be strongly normalizing, i.e., to belong to $N_R^s(R[X, \alpha])$. Indeed, since $Xr = \alpha(r)X$ for any $r \in R$, we also have $X^n r = \alpha^n(r)X^n$. Hence $X^n R = RX^n$, i.e., $X^n \in N_R(R[X, \alpha])$. On the other hand, we also have

$$X^n r = \alpha^n(r)X^n = a^{-n}ra^n X^n = a^{-n}rX^n a^n \in RrX^n R,$$

as $Xa = \alpha(a)X = aX$. By symmetry, we also have $rX^n \in RX^n rR$, so $RrX^n R = RX^n rR$. It follows that the conditions of (3.4.2) below are satisfied, hence $X^n \in N_R^s(R[X, \alpha])$, indeed. $\qquad\square$

(3.4) Lemma. *Let M be an R-bimodule and let $m \in M$, then the following assertions are equivalent:*

(3.4.1) $m \in N_R^s(M)$;

(3.4.2) $m \in N_R(M)$ *and* $RmrR = RrmR$ *for all* $r \in R$.

Proof. Assume that (1) holds, then, in particular, $Rm = mR$, so we certainly have $m \in N_R(M)$. On the other hand, if $r \in R$, then, applying the definition of $m \in N_R^s(M)$ to the twosided ideal RrR we obtain

$$RrmR = (RrR)m = m(RrR) = RmrR.$$

Conversely, assume (2) and let I be an arbitrary twosided ideal of R. If $i \in I$, then we obtain

$$im \in RimR = RmiR = mRiR \subseteq mI.$$

So, $Im \subseteq mI$, whence equality by symmetry.   □

Exactly as in (1.3), one proves:

(3.5) Lemma. If M is a strongly normalizing R-bimodule, then $IM = MI$ for any twosided ideal I of R.

Let us also point out the following useful result:

(3.6) Lemma. Let M be an R-bimodule and let $m \in N_R^s(M)$, then the left annihilator $Ann_R^\ell(m)$ is a twosided ideal of R and coincides with the right annihilator $Ann_R^r(m)$.

Proof. Of course, the second statement follows from the first one. If $r \in Ann_R^\ell(m)$ then $rm = 0$, so $RmrR = RrmR = 0$. Hence $mr = 0$ and $r \in Ann_R^r(m)$. So, $Ann_R^\ell(m) \subseteq Ann_R^r(m)$, whence equality by symmetry.   □

(3.7) We have already recalled, cf. (2.2), that for any R-bimodule M, an element $m \in M$ is automorphic if there exists some automorphism φ_m of R, with $rm = m\varphi_m(r)$ for all $r \in R$. If φ_m also has the property that $RrR = R\varphi_m(r)R$ for all $r \in R$, then we call m *strongly automorphic*.

Obviously, strongly automorphic elements are strongly normalizing. So, every R-bimodule $M = \sum Rm_i$, where the m_i are strongly automorphic is, of course, strongly normalizing. We will call these R-bimodules *strongly automorphic* We will see in a moment that, up to torsion, essentially all strongly normalizing R-bimodules are of this type.

To prove this, first recall that a left R-module N is said to be *absolutely torsionfree* if for any $n \in N$ and any twosided ideal I of R, it follows from $In = 0$ that $I = 0$ or $n = 0$.

For example, if $R \subseteq S$ is a strongly normalizing extension with the property that S is prime, then this extension makes S into an absolutely torsionfree R-module. Indeed, if I is a twosided ideal of R and if $s \in S$, then from $Is = 0$ we deduce $ISs = SIs = 0$, so $I = 0$ or $s = 0$, as S is prime.

We are now ready to prove:

(3.8) Proposition. *Let M be an absolutely torsionfree R-bimodule, then M is strongly normalizing if and only if M is strongly automorphic.*

Proof. We just mentioned that any strongly automorphic R-bimodule is strongly normalizing.

Conversely, assume M to be a strongly normalizing R-bimodule and let $m \in N_R^s(M)$. It is then clear that $Ann_R^\ell(m) = Ann_R^r(m) = 0$. Indeed, this follows immediately from the fact that M is absolutely torsionfree and that $Ann_R^r(m)$ is a twosided ideal by the previous lemma. Now, for any $r \in R$, there exists some r' with the property that $rm = mr'$ and, by the previous remarks, r' is unique as such. If we define φ_m by putting $\varphi_m(r) = r'$, then it is easy to see that this actually defines an automorphism of R. Moreover, from

$$m(R\varphi(r)R) = Rm\varphi(r)R = RrmR = RmrR = m(RrR),$$

it follows that $R\varphi(r)R = RrR$, since $Ann_R^r(m) = 0$. This proves that m is strongly automorphic, whence the assertion. $\square$

Let us now briefly indicate how strongly normalizing extension behave with respect to composition and quotients.

(3.9) Proposition. *If $\varphi : R \to S$ is a strongly normalizing extension and $\psi : S \to T$ a centralizing extension, then the composition $\psi\varphi : R \to T$ is strongly normalizing.*

Proof. By assumption, $S = \sum_i \varphi(R)s_i$ for certain $s_i \in N_R^s(S)$, i.e., with the property that $\varphi(R)s_i\varphi(r)\varphi(R) = \varphi(R)\varphi(r)s_i\varphi(R)$ for any index i and any $r \in R$. On the other hand, $T = \sum_j \psi(S)t_j$ for certain $t_j \in S^R$, i.e., with the property that $\psi(s)t_j = t_j\psi(s)$ for every index j and any $s \in S$.

It is then clear that $T = \sum_{i,j} \psi\varphi(R)\psi(s_i)t_j$ and since one easily verifies that $\psi(s_i)t_j \in N_R^s(T)$ for any pair of indices i, j, this proves that $\psi\varphi : R \to T$ is strongly normalizing, indeed. $\square$

Of course, the same proof also yields:

(3.10) Corollary. If $\varphi : R \to S$ is a strongly normalizing extension and M a centralizing S-bimodule, then M is a strongly normalizing R-bimodule, when endowed with the R-bimodule structure induced through φ.

As an immediate corollary, let us mention:

(3.11) Corollary. If $\varphi : R \to S$ is a strongly normalizing extension, then so is

$$R \xrightarrow{\varphi} S \to S/I$$

for any twosided ideal I of S.

It is also fairly straightforward to prove:

(3.12) Lemma. If $\varphi : R \to S$ is a strongly normalizing extension, then so is the induced map $\overline{\varphi} : R/\varphi^{-1}(I) \hookrightarrow S/I$ for any twosided ideal I of S.

In particular, this yields (for $I = 0$):

(3.13) Corollary. If $\varphi : R \to S$ is a strongly normalizing extension, then so is the induced homomorphism $\overline{\varphi} : R/Ker(\varphi) \hookrightarrow S$.

This last property, combined with the well-known features of centralizing and, in particular, of surjective morphisms, permits us to restrict to injective morphisms in most of what follows. We leave it to the reader to reformulate the results below, in the general case.

The power of strongly normalizing extensions partially resides in the fact that they behave nicely with respect to prime ideals.

Let us start with the following fundamental fact:

(3.14) Proposition. If $\varphi : R \to S$ is a strongly normalizing extension, then $\varphi^{-1}(Q)$ is a prime ideal of R, for any prime ideal Q of S.

Proof. Indeed, as $SI = IS$ for any twosided ideal I of R, cf. (3.6), this may be proved exactly as in the centralizing case, cf. (1.5.2). $\square$

From (2.4), it now follows:

(3.15) Corollary. Let $R \subseteq S = \sum_{i=1}^{n} Rs_i$ be a finite strongly normalizing extension, then:

(3.15.1) (Lying over) for every prime ideal P of R, there exists at least one and at most n prime ideals Q of S with $Q \cap R = P$;

(3.15.2) (Going up) for every prime ideal Q of S and for every prime ideal P' of R with $Q \cap R \subseteq P'$, there exists a prime ideal Q' of S with $Q' \cap R = P'$ and $Q \subseteq Q'$;

(3.15.3) (Incomparability) for any pair of prime ideals $Q \subseteq Q'$ of S with $Q \cap R = Q' \cap R$, we have $Q = Q'$.

As an easy consequence of this, let us mention the following technical result (to be compared with (3.8)), which we will need below:

(3.16) Lemma. If $R \subseteq S$ is a finite strongly normalizing extension, then for any $s \in N_R^s(S)$, we have $Ann_R^\ell(s) \subseteq rad(R)$.

Proof. It has already been pointed out that if S is prime, then for any $s \in N_R^s(S)$ we have $Ann_R^\ell(s) = 0$. In the general case, assume $P \in Spec(R)$, then for some $Q \in Spec(S)$ we have $Q \cap R = P$ and $R \subseteq S$ induces a strongly normalizing extension $R/P \subseteq S/Q$. For any

$s \in N_R^s(S)$, we have $\overline{s} \in N_{R/P}^s(S/Q)$, so, by the foregoing remarks, $Ann_{R/P}^\ell(\overline{s}) = 0$, whence $Ann_R^\ell(s) \subseteq P$. As this holds for any $P \in Spec(R)$, this proves the assertion. $\qquad\square$

4.   Inducing radicals

(4.1) Let us consider a pair of rings (R, S) linked by adjoint functors

$$R-\mathbf{mod} \underset{G}{\overset{F}{\rightleftarrows}} S-\mathbf{mod}.$$

If G is exact and commutes with direct sums, then for any radical σ in $R-\mathbf{mod}$, we may define a radical $G(\sigma)$ in $S-\mathbf{mod}$ by letting $\mathcal{T}_{G(\sigma)}$ consist of all left S-modules N with the property that $G(N) \in \mathcal{T}_\sigma$. Dually, if F is exact and commutes with direct sums, then for every radical τ in $S-\mathbf{mod}$, we define the radical $F(\tau)$ in $R-\mathbf{mod}$ by letting $\mathcal{T}_{F(\tau)}$ consist of all left R-modules M with the property that $F(M) \in \mathcal{T}_\tau$.

(4.2) As a particular case, consider a ring homomorphism $\varphi : R \to S$, then we may associate to it a pair of adjoint functors

$$R-\mathbf{mod} \underset{\varphi_*}{\overset{\varphi^*}{\rightleftarrows}} S-\mathbf{mod},$$

where φ_* is the restriction of scalars functor, which associates to any $N \in S-\mathbf{mod}$ the left R-module ${}_R N$ defined by putting $rn = \varphi(r)n$ for any $r \in R$ and $n \in N$, and where φ^* is the extension of scalars functor which associates to $M \in R-\mathbf{mod}$ the left S-module $S \otimes_R M$. It is clear that φ_* is exact and commutes with direct sums. So, for any radical σ in $R-\mathbf{mod}$, there exists a radical $\bar{\sigma} = \varphi_*\sigma$ in $S-\mathbf{mod}$, with

$$\mathcal{T}_{\varphi_*\sigma} = \mathcal{T}_{\bar{\sigma}} = \{N \in S-\mathbf{mod};\ {}_R N \in \mathcal{T}_\sigma\}.$$

In other words, a left S-module N is $\bar{\sigma}$-torsion, if and only if it is σ-torsion when considered as a left R-module through φ.

On the other hand, if φ is flat, for example, then φ^* is exact and in this case for any radical τ in $S-\mathbf{mod}$, we may define $\varphi^*\tau$ in $R-\mathbf{mod}$ by letting

$$\mathcal{T}_{\varphi^*\tau} = \{M \in R-\mathbf{mod};\ S \otimes_R M \in \mathcal{T}_\tau\}.$$

This definition does not yield a radical in the general case, i.e., when φ is not necessarily flat.

If $\varphi : R \to S$ is a ring homomorphism, then we have already seen that the torsion classes $\mathcal{T}_\sigma$ and $\mathcal{T}_{\overline{\sigma}}$ are linked in an obvious way. In general, it is not true that the torsion free class $\mathcal{F}_{\overline{\sigma}}$ coincides with

$$\{N \in S-\mathbf{mod};\ {}_R N \in \mathcal{F}_\sigma\}.$$

When this condition does hold, then we will say that the radical σ is *compatible* with the ring homomorphism $\varphi : R \to S$.

Actually, one may prove the following result:

(4.3) Proposition. [33] *Let $\varphi : R \to S$ be a ring homomorphism and let σ be a radical. Then the following assertions are equivalent:*

(4.3.1) *σ is compatible with φ, i.e., a left S-module N is $\overline{\sigma}$-torsionfree if and only if ${}_R N$ is σ-torsionfree;*

(4.3.2) *if M is σ-torsionfree, then so is $Hom_R(S, M)$;*

(4.3.3) *if M is σ-torsion, then so is $S \otimes_R M$;*

(4.3.4) *the Gabriel filter $\mathcal{L}(\overline{\sigma})$ consists of all left S-ideals L with the property that $\varphi^{-1}(L) \in \mathcal{L}(\sigma)$;*

(4.3.5) *for any $N \in S-\mathbf{mod}$, we have $\overline{\sigma}N = \sigma({}_R N)$.*

Proof. Let us first prove that (1) implies (2). So, let $M \in \mathcal{F}_\sigma$ and let us show that $Hom_R(S, M) \in \mathcal{F}_\sigma$ as well. If we put $N = \sigma Hom_R(S, M)$, then

$$Hom_S(N, Hom_R(S, M)) = Hom_R(N, M) = 0,$$

so $N = 0$, indeed.

Next, to show that (2) implies (3), let $M \in \mathcal{T}_\sigma$ and pick an arbitrary $F \in \mathcal{F}_\sigma$. Then, as $Hom_R(S, F) \in \mathcal{F}_\sigma$ by (2),

$$\begin{aligned} Hom_R(S \otimes_R M, F) &= Hom_S(S \otimes_R M, Hom_R(S, F)) \\ &= Hom_R(M, Hom_R(S, F)) = 0, \end{aligned}$$

so $S \otimes_R M \in \mathcal{T}_\sigma$, as this holds for any $F \in \mathcal{F}_\sigma$.

To show that (3) implies (1), let $N \in S-\mathbf{mod}$ have the property that $_R N \in \mathcal{F}_\sigma$, then clearly $N \in \mathcal{F}_{\bar{\sigma}}$. Conversely, if $N \in \mathcal{F}_{\bar{\sigma}}$, then for any $M \in \mathcal{T}_\sigma$, we have

$$Hom_R(M, N) = Hom_S(S \otimes_R M, N) = 0,$$

as $S \otimes_R M \in \mathcal{T}_{\bar{\sigma}}$ by (3). So, $N \in \mathcal{F}_\sigma$, indeed.

Next, to show that (1) and (4) are equivalent, first assume (1) and consider $L \in \mathcal{L}(\bar{\sigma})$, then $R/\varphi^{-1}(L) \hookrightarrow S/L$, so $\varphi^{-1}(L) \in \mathcal{L}(\sigma)$ as $_R(S/L) \in \mathcal{T}_\sigma$. Conversely, if $\varphi^{-1}(L) \in \mathcal{L}(\sigma)$ and if $L \notin \mathcal{L}(\bar{\sigma})$, then there exists $K \supseteq L$ with $S/K \in \mathcal{F}_{\bar{\sigma}}$. As $\varphi^{-1}(L) \subseteq \varphi^{-1}(K)$, clearly $\varphi^{-1}(K) \in \mathcal{L}(\sigma)$. By (1) it follows that $_R(S/K) \in \mathcal{F}_\sigma$, hence $R/\varphi^{-1}(K) \in \mathcal{F}_\sigma$ as well – a contradiction.

On the other hand, if $N \in S-\mathbf{mod}$ has the property that $_R N \in \mathcal{F}_\sigma$, then obviously $N \in \mathcal{F}_{\bar{\sigma}}$. Conversely, assume (4) and let $N \in \mathcal{F}_{\bar{\sigma}}$. Let $n \in \sigma(_R N)$ and put $L = Ann_S^\ell(n)$, then $\varphi^{-1}(L) = Ann_R^\ell(n) \in \mathcal{L}(\sigma)$. From (4), it follows that $L \in \mathcal{L}(\bar{\sigma})$, hence $n \in \sigma N = 0$. So, $_R N \in \mathcal{F}_\sigma$, indeed.

Finally, to show that (1) implies (5), first note that $\bar{\sigma} N \subseteq \sigma(_R N)$ for any $N \in S-\mathbf{mod}$. Conversely, as $N/\bar{\sigma} N \in \mathcal{F}_{\bar{\sigma}}$, it follows from (1) that $_R(N/\bar{\sigma} N) \in \mathcal{F}_\sigma$, hence $\sigma(_R N) \subseteq \bar{\sigma} N$, whence equality.

Finally, assume (5) and let us prove (4). Clearly, if $L \in \mathcal{L}(\bar{\sigma})$, then $\varphi^{-1}(L) \in \mathcal{L}(\sigma)$, as $R/\varphi^{-1}(L) \hookrightarrow S/L$. Conversely, let L be a left ideal of S with the property that $\varphi^{-1}(L) \in \mathcal{L}(\sigma)$. Clearly $\bar{1} \in \sigma(_R(S/L))$, as it is annihilated by $\varphi^{-1}(L)$. Hence $\bar{1} \in \bar{\sigma}(S/L)$, i.e., there exists some $K \in \mathcal{L}(\bar{\sigma})$, with $K\bar{1} = \bar{0}$, i.e., $K \subseteq L$. So, $L \in \mathcal{L}(\bar{\sigma})$, which proves the assertion. $\qquad\square$

(4.4) Proposition. [33] *Let $\varphi : R \to S$ be a flat ring homomorphism, let τ be a radical in $S-\mathbf{mod}$ and $\sigma = \varphi^* \tau$. Consider the following assertions:*

(4.4.1) $\tau \leq \varphi_* \sigma$;

(4.4.2) σ is compatible with φ;

(4.4.3) $\sigma = \varphi^* \varphi_* \sigma$.

Then we have the following implications: $(1) \Rightarrow (2) \Rightarrow (3)$.

Proof. That (1) implies (2) is obvious. Indeed, if $M \in R-\textbf{mod}$, then $M \in \mathcal{T}_\sigma$ if and only if $M \in \mathcal{T}_{\varphi^* \tau}$, i.e., if and only if $S \otimes_R M \in \mathcal{T}_\tau$, by definition. But, as $\tau \leq \phi_* \sigma$, this implies $S \otimes_R M \in \mathcal{T}_{\varphi_* \sigma}$, hence σ is compatible with φ, indeed.

Next, if σ is compatible with φ and if $M \in \mathcal{T}_\sigma$, then $S \otimes_R M \in \mathcal{T}_{\varphi_* \sigma}$, i.e., $M \in \mathcal{T}_{\varphi^* \varphi_* \sigma}$, so $\sigma \leq \varphi^* \varphi_* \sigma$. So, to prove that (2) implies (3), it remains to prove the other inequality $\sigma \geq \varphi^* \varphi_* \sigma$.

Let $L \in \mathcal{L}(\varphi_* \sigma)$, then $\varphi^{-1}(L) \in \mathcal{L}(\sigma = \varphi^* \tau)$, by the foregoing result. So $S/S\varphi^{-1}(L) = S \otimes_R R/\varphi^{-1}(L) \in \mathcal{T}_\tau$, hence $S\varphi^{-1}(L) \in \mathcal{L}(\tau)$. As $S\varphi^{-1}(L) \subseteq L$, this shows that $L \in \mathcal{L}(\tau)$ as well, which proves that $\varphi_* \sigma \leq \tau$, whence (a fortiori) $\varphi^* \varphi_* \sigma \leq \varphi^* \tau = \sigma$.   □

Let us now investigate how radicals behave with respect to finite, centralizing, normalizing and strongly normalizing extensions.

We start with an easy result about finite extensions.

(4.5) Lemma. Let M be an R-bimodule, which is finitely generated as a right R-module. Then the following assertions are equivalent:

(4.5.1) $M \in \mathcal{T}_\sigma$;

(4.5.2) $Ann^\ell_R(M) \in \mathcal{L}(\sigma)$;

(4.5.3) there exists some $I \in \mathcal{L}^2(\sigma)$ with $IM = 0$.

Proof. The implications $(2) \Rightarrow (3) \Rightarrow (1)$ being obvious, let us show that (1) implies (2) to finish the proof. Since M is finitely generated as a right R-module, we may put $M = \sum_{i=1}^n m_i R$ for some $m_i \in M$. As M is σ-torsion, for each $1 \leq i \leq n$, the annihilator $Ann^\ell_R(m_i)$ belongs to $\mathcal{L}(\sigma)$, hence so does $I = \bigcap_{i=1}^n Ann^\ell_R(m_i)$. Since obviously $I \subseteq Ann^\ell_R(M)$, this proves that $Ann^\ell_R(M) \in \mathcal{L}(\sigma)$, indeed.   □

As a Corollary, we obtain:

(4.6) Proposition. *If $R \subseteq S$ is a finite ring homomorphism, then for any radical σ we have*

$$\mathcal{L}^2(\overline{\sigma}) = \{L \trianglelefteq S; \; L \cap R \in \mathcal{L}^2(\sigma)\}.$$

Proof. First, assume $L \in \mathcal{L}^2(\overline{\sigma})$, then $S/L \in \mathcal{T}_{\overline{\sigma}}$ or, equivalently, $_R(S/L) \in \mathcal{T}_\sigma$. As our assumption implies S/L to be a finitely generated right R-module, we may apply the previous result. So, there exists some $I \in \mathcal{L}^2(\sigma)$ with the property that $I(S/L) = 0$, i.e., with $IS \leq L$. But then

$$I \leq IS \cap R \leq L \cap R,$$

which implies that $L \cap R \in \mathcal{L}^2(\sigma)$, indeed.

Conversely, let L be a twosided ideal of S with the property that $L \cap R \in \mathcal{L}^2(\sigma)$. Then $R/L \cap R \in \mathcal{T}_\sigma$, and since this is a left and right finitely generated R-bimodule, there exists $I \in \mathcal{L}^2(\sigma)$ with $IR \subseteq L \cap R$. But then $IS \subseteq (L \cap R)S \subseteq L$, hence $_R(S/L) \in \mathcal{T}_\sigma$. This is equivalent to $S/L \in \mathcal{T}_{\overline{\sigma}}$, i.e., $L \in \mathcal{L}^2(\sigma)$, indeed. $\square$

(4.7) Consider a ring extensions $R \to S$, which we assume to be injective, for notational convenience, and let $s \in N_R(S)$. Then for any left ideal L of R, we define the *symmetrization* L_s^* of L with respect to s by

$$L_s^* = \{r \in R; \; rs \in sL\}.$$

For every $s \in N_R(S)$, it is clear that L_s^* is a left ideal of R with the property that $L_s^* s \subseteq sL$.

(4.8) Proposition. [55] *Let $\varphi : R \to S$ be a ring extension such that S is generated over R by a subset $N \subseteq N_R(S)$ and let σ be a radical in $R-\mathbf{mod}$ with the property that $\mathcal{L}(\sigma)$ has a filter basis $\mathcal{L}$ such that for every $L \in \mathcal{L}$ and every $s \in N$ we have $L_s^* \in \mathcal{L}(\sigma)$, then σ is compatible with φ.*

Proof. Let us verify (4.3.3). So, let $M \in \mathcal{T}_\sigma$ and consider an arbitrary element $x \in S \otimes_R M$, then x will be of the form $x = \sum s_i \otimes m_i$ with

$s_i \in S$ and $m_i \in M$. We may assume without loss of generality φ to be an inclusion, so each s_i will be of the form $\sum z_\alpha r_{\alpha i}$ for some $r_{\alpha i} \in R$ and $z_\alpha \in N$. We thus obtain that

$$x = \sum_i s_i \otimes m_i = \sum_\alpha \sum_i z_\alpha r_{\alpha i} \otimes m_i = \sum_\alpha z_\alpha \otimes \sum_i r_{\alpha i} m_i.$$

In other words, $x = \sum_{\alpha \in A} z_\alpha \otimes n_\alpha$ for some finite index set A and some $n_\alpha \in M$. For each $\alpha \in A$, the element n_α is σ-torsion, so $Ann_R^\ell(n_\alpha) \in \mathcal{L}(\sigma)$. Hence $J = \bigcap_{\alpha \in A} Ann_R^\ell(n_\alpha) \in \mathcal{L}(\sigma)$ as well. As $\mathcal{L}$ is a filter basis for $\mathcal{L}(\sigma)$, there exists some $I \in \mathcal{L}$ with $I \subseteq J$. Applying our assumption, it follows that

$$I^* = \bigcap_{\alpha \in A} I_{z_\alpha}^* \in \mathcal{L}(\sigma),$$

and we obtain:

$$I^* x = I^* \left(\sum_{\alpha \in A} z_\alpha \otimes n_\alpha \right) = \sum_{\alpha \in A} I^* z_\alpha \otimes n_\alpha \subseteq \sum_{\alpha \in A} I_{z_\alpha}^* z_\alpha \otimes n_\alpha$$

$$\subseteq \sum_{\alpha \in A} z_\alpha I \otimes n_\alpha \subseteq \sum_{\alpha \in A} z_\alpha J \otimes n_\alpha = \sum_{\alpha \in A} z_\alpha \otimes J n_\alpha = 0.$$

So, $x \in \sigma(S \otimes_R M)$ and it follows that $S \otimes_R M \in \mathcal{T}_\sigma$, indeed, as x is arbitrary in $S \otimes_R M$. $\qquad\square$

As a particular case, let us mention:

(4.9) Proposition. *If $\varphi : R \to S$ is a strongly normalizing extension, then every symmetric radical in $R-\mathbf{mod}$ is compatible with φ.*

Proof. If σ is symmetric in $R-\mathbf{mod}$, then we may apply the previous result with $N = N_R^s(S) \subseteq N_R(S)$ and $\mathcal{L} = \mathcal{L}^2(\sigma)$. Indeed, for every $I \in \mathcal{L}^2(\sigma)$ and every $s \in N_R^s(S)$, we have $Is = sI$, so $I_s^* = I$. $\qquad\square$

(4.10) Proposition. *If $\varphi : R \to S$ is a strongly normalizing extension, then for every symmetric radical σ, the induced radical $\overline{\sigma}$ in $S-\mathbf{mod}$ is symmetric as well.*

Proof. Let $I \in \mathcal{L}(\overline{\sigma})$, then by the previous result, $\varphi^{-1}(I) \in \mathcal{L}(\sigma)$. So there exists $J \in \mathcal{L}^2(\sigma)$ with $J \subseteq \varphi^{-1}(I)$. Clearly, $S\varphi(J)$ is a twosided ideal of S and moreover $J \subseteq \varphi^{-1}(S\varphi(J))$, so $\varphi^{-1}(S\varphi(J)) \in \mathcal{L}(\sigma)$. It follows that $S\varphi(J) \in \mathcal{L}^2(\overline{\sigma})$. Finally, as $J \subseteq \varphi^{-1}(I)$, we obtain $\varphi(J) \subseteq I$, proving that $S\varphi(J) \subseteq I$, indeed. $\square$

(4.11) Proposition. *If* $\varphi : R \to S$ *is a centralizing extension, then every radical in* $R-\mathbf{mod}$ *is compatible with* φ.

Proof. It suffices to apply (4.8) to $\mathcal{L} = \mathcal{L}(\sigma)$ and $N = S^R$. $\square$

(4.12) Corollary. *If* $\varphi : R \to S$ *is a surjective ring morphism, then every radical in* $R-\mathbf{mod}$ *is compatible with* φ.

Proof. It suffices to note that any surjective morphism is centralizing. $\square$

Let us conclude this Section by briefly studying the behaviour of $\mathcal{K}(\sigma)$ with respect to extensions.

(4.13) Proposition. *Let* $i : R \hookrightarrow S$ *be a finite normalizing extension and assume* R *to be left noetherian. Then, for any radical* σ *in* $R-\mathbf{mod}$, *we have*

$$\mathcal{K}(\overline{\sigma}) = \{Q \in Spec(S); \; (^a i)(Q) \cap \mathcal{K}(\sigma) \neq \varnothing\} = (^a i)^{-1}(\mathcal{K}(\sigma)),$$

where $(^a i)(Q)$ *is the set of all prime ideals of* R *which are minimal over* $Q \cap R$.

Proof. Since R is assumed to be left noetherian, so is S and we thus obtain that $\mathcal{K}(\sigma) = Spec(R) \setminus \mathcal{L}^2(\sigma)$ resp. $\mathcal{K}(\overline{\sigma}) = Spec(S) \setminus \mathcal{L}^2(\overline{\sigma})$. It thus follows that

$$\mathcal{K}(\overline{\sigma}) = \{Q \in Spec(S); \; Q \cap R \notin \mathcal{L}(\sigma)\}.$$

But, since $Q \cap R = \bigcap\{P \in Spec(R); \; P \in (^a i)(Q)\}$ and since this is a finite intersection, obviously $Q \cap R \notin \mathcal{L}(\sigma)$ if and only if $(^a i)(Q) \cap \mathcal{K}(\sigma) \neq \varnothing$. $\square$

(4.14) Proposition. Let $R \subseteq S$ be a strongly normalizing extension. Then, for any symmetric radical σ in $R-\mathbf{mod}$, we have

$$\mathcal{K}(\overline{\sigma}) = \{Q \in Spec(S); \ Q \cap R \in \mathcal{K}(\sigma)\}.$$

Proof. Since (4.10) implies that $\overline{\sigma}$ is symmetric, as σ is, it follows that $\mathcal{K}(\sigma) = Spec(R) \setminus \mathcal{L}(\sigma)$ and that $\mathcal{K}(\overline{\sigma}) = Spec(S) \setminus \mathcal{L}(\overline{\sigma})$. The result now follows exactly as in the previous Proposition, using the fact that $Q \cap R \in Spec(R)$ for any $Q \in Spec(S)$. $\qquad\square$

5. Inducing localization functors

(5.1) As in the previous section, let us consider a ring homomorphism $\varphi : R \to S$ and a radical σ in $R-\mathbf{mod}$. Let us denote by $\overline{\sigma}$ the induced radical $\varphi_*\sigma$ in $S-\mathbf{mod}$, then it is fairly obvious that we want to relate Q_σ, the localization functor in $R-\mathbf{mod}$ with respect to σ and $Q_{\overline{\sigma}}$, the localization functor in $S-\mathbf{mod}$ with respect to $\overline{\sigma}$. On the other hand, we are also interested in trying to extend φ to a ring homomorphism $\overline{\varphi} : Q_\sigma(R) \to Q_{\overline{\sigma}}(S)$. It goes without saying that the two questions are closely related.

Let us start with the following positive result:

(5.2) Proposition. *Let $\varphi : R \to S$ be a centralizing extension resp. a strongly normalizing extension and let σ be a radical resp. a symmetric radical in $R-\mathbf{mod}$, then for any left S-module M, the following assertions are equivalent:*

(5.2.1) $Q_\sigma(M) \in S-\mathbf{mod}$, *its S-module structure extending that of M;*

(5.2.2) *there exists a canonical isomorphism $Q_\sigma(M) = Q_{\overline{\sigma}}(M)$ of left R-modules;*

(5.2.3) *there exists a radical $\tau \geq \sigma$ in $R-\mathbf{mod}$ with the property that $\sigma M = \tau M$ and that $Q_\tau(M) \in S-\mathbf{mod}$, its S-module structure extending that of M.*

Proof. Let us first verify that (1) implies (2) in the centralizing case. To start with, let us show that $Q_\sigma(M)$ is faithfully $\overline{\sigma}$-injective. Let $L \in \mathcal{L}(\overline{\sigma})$ and consider a left S-linear map $f : L \to Q_\sigma(M)$. Then, since $S/L \in \mathcal{T}_\sigma$, it follows that f extends to a unique left R-linear map $\overline{f} : S \to Q_\sigma(M)$. Let us show that $\overline{f}$ is also left S-linear. Let $q, s \in S$, and put $f_s(q) = \overline{f}(sq) - s\overline{f}(q)$. Clearly, $f_s(q) = 0$ if $s = \varphi(r)$ for some $r \in R$. On the other hand, if $s \in S^R$ then, obviously, $f_s(rq) = rf_s(q)$, so $f_s : S \to Q_\sigma(M)$ is left R-linear. Since $f_s|L = 0$, it factorizes through

$\overline{f}_s : S/L \to Q_\sigma(M)$. But, $S/L \in \mathcal{T}_\sigma$, so $\overline{f}_s = 0$, hence $f_s = 0$. It follows that $\overline{f}(sq) = s\overline{f}(q)$ for $s \in \varphi(R)$ and $s \in S^R$, hence $\overline{f}(sq) = s\overline{f}(q)$ for all s. So $\overline{f}$ is S-linear, indeed. Finally, since the kernel and cokernel of $j : M \to Q_\sigma(M)$ are σ-torsion, hence $\overline{\sigma}$-torsion left S-modules, this shows that $Q_\sigma(M) = Q_{\overline{\sigma}}(M)$ as left R-modules, indeed.

In the strongly normalizing case (with σ symmetric), one proceeds in a similar way. There, one starts from a left S-linear map $f : L \to Q_\sigma(M)$, where L may now be taken within $\mathcal{L}^2(\overline{\sigma})$. The only point left to verify, in this case, is whether the left R-linear extension $\overline{f} : S \to Q_\sigma(M)$ satisfies $\overline{f}(sq) = s\overline{f}(q)$, for $q \in S$ and $s \in N_R^s(S)$. Let $I = L \cap R \in \mathcal{L}^2(\sigma)$ and pick $j \in I$, then $js = sj'$ for some $j' \in I$. So, we obtain

$$\begin{aligned} j\overline{f}(sq) &= \overline{f}(jsq) = \overline{f}(sj'q) = f(sj'q) = \\ &= sf(j'q) = s\overline{f}(j'q) = sj'\overline{f}(q) = \\ &= js\overline{f}(q) \end{aligned}$$

It follows that $I(\overline{f}(sq) - s\overline{f}(q)) = 0$, hence that $\overline{f}(sq) - s\overline{f}(q) \in \sigma Q_\sigma(M) = 0$, i.e., $\overline{f}(sq) = s\overline{f}(q)$, indeed.

The implications $(2) \Rightarrow (1)$ and $(1) \Rightarrow (3)$ being trivial, let us conclude by showing that (3) implies (1).

Let us first work in the centralizing case and assume $q \in Q_\sigma(M) \subseteq Q_\tau(M)$. Then, for any $s \in S$ we have $sq \in Q_\tau(M)$. Let us show that $sq \in Q_\sigma(M)$ and leave the verification that this yields a well-defined left S-module structure on $Q_\sigma(M)$ to the reader. Let $s = \sum r_\alpha z_\alpha =: \sum \varphi(r_\alpha) z_\alpha$ for some $r_\alpha \in R$ and $z_\alpha \in S^R$. We may clearly assume M to be σ-torsionfree, so $Iq \subseteq M$ for some $I \in \mathcal{L}(\sigma)$. Since

$$(I : r_\alpha) = \{r \in R;\ rr_\alpha \in I\} \in \mathcal{L}(\sigma),$$

we also have that $J = \bigcap_{\alpha \in A}(I : r_\alpha) \in \mathcal{L}(\sigma)$. Let $j \in J$, then, for all α, we get $jr_\alpha = i_\alpha \in I$ for some $i_\alpha \in I$, so

$$j(sq) = j(\sum_\alpha r_\alpha z_\alpha q) = \sum_\alpha i_\alpha z_\alpha q = \sum_\alpha z_\alpha(i_\alpha q) \in M.$$

Hence, $J(sq) \subseteq M$ and $sq \in Q_\sigma(M)$, indeed.

In the strongly normalizing case (with σ symmetric), the z_α now have to be taken in $N_R^s(S)$ and I may be assumed to belong to $\mathcal{L}(\sigma)$. For any $j \in J = \bigcap_\alpha (I : r_\alpha)$, we again have $jr_\alpha = i_\alpha \in I$ and for each α we may find $i'_\alpha \in I$ with $i_\alpha z_\alpha = z_\alpha i'_\alpha$ (as $Iz_\alpha = z_\alpha I$, since $z_\alpha \in N_R^s(S)$). So,

$$j(sq) = j(\sum_\alpha r_\alpha z_\alpha q) = \sum_\alpha i_\alpha z_\alpha q = \sum_\alpha z_\alpha (i'_\alpha q) \in M,$$

which again implies that $J(sq) \subseteq M$ and $sq \in Q_\sigma(M)$. This finishes the proof.   □

Applying the previous result to the case $M = S$ yields the following generalization of [1, (1.7.)]:

(5.3) Corollary. *Let $\varphi : R \to S$ be a centralizing resp. strongly normalizing extension and let σ be a radical resp. a symmetric radical in $R-$mod, then the following assertions are equivalent:*

(5.3.1) *$Q_\sigma(S) \in S-$mod, its S-module structure extending that of S;*

(5.3.2) *there exists a canonical isomorphism $Q_\sigma(S) = Q_{\bar{\sigma}}(S)$ of left R-modules;*

(5.3.3) *there exists a radical $\tau \geq \sigma$ in $R-$mod with the property that $\sigma S = \tau S$ and $Q_\tau(S) \in S-$mod, its S-structure extending that of S;*

(5.3.4) *$Q_\sigma(S)$ has a ring structure, making the canonical morphism $S \to Q_\sigma(S)$ into a ring morphism.*

Proof. Only the equivalence of (4) and the previous statements remains to be verified. Of course, if $Q_\sigma(S)$ has a ring structure, extending that of S, then (1) holds. On the other hand, (4) trivially follows from (2).   □

As a first example, let us mention:

(5.4) Proposition. Let $R \subseteq S$ be a strongly normalizing extension of rings and assume S (and hence R) to be prime. Let σ be a symmetric radical in $R-\mathbf{mod}$ with the property that any $I \in \mathcal{L}(\sigma)$ intersects the center $Z(R)$ of R non-trivially. Then $Q_{\overline{\sigma}}(S)$ is σ-closed.

Proof. Since $\sigma(_R Q_{\overline{\sigma}}(S)) = {}_R(\overline{\sigma} Q_{\overline{\sigma}}(S)) = 0$, it suffices to show that the canonical map

$$Hom_R(R, {}_R Q_{\overline{\sigma}}(S)) \to Hom_R(I, {}_R Q_{\overline{\sigma}}(S))$$

is surjective for all $I \in \mathcal{L}(\sigma)$. So, choose $I \in \mathcal{L}(\sigma)$ and let $f : I \to Q_{\overline{\sigma}}(S)$ be left R-linear. Define a left S-linear morphism $g : S\varphi(I) \to Q_{\overline{\sigma}}(S)$ by putting $g(\sum s_\alpha j_\alpha) = \sum s_\alpha f(j_\alpha)$, where $s_\alpha \in S$ and $j_\alpha \in I$. Let us first verify that g is well-defined. For this, let $0 \neq c \in I \cap Z(R)$. Then $c \in N(S)$, so, for every index α, there exists some $s'_\alpha \in S$, such that $cs_\alpha = s'_\alpha c$. Moreover, if $\sum_\alpha s_\alpha j_\alpha = 0$, then $\sum_\alpha s'_\alpha j_\alpha = 0$, since

$$0 = c(\sum_\alpha s_\alpha j_\alpha) = \sum_\alpha s'_\alpha c j_\alpha = (\sum_\alpha s'_\alpha j_\alpha)c$$

and since c is regular in S, as S is prime and $c \in N(S)$. Now, if $\sum_\alpha s_\alpha j_\alpha = 0$, then we have

$$c(\sum_\alpha s_\alpha f(j_\alpha)) = \sum_\alpha s'_\alpha c f(j_\alpha) = \sum_\alpha s'_\alpha f(c j_\alpha)$$
$$= \sum_\alpha s'_\alpha f(j_\alpha c) = (\sum_\alpha s'_\alpha j_\alpha)f(c) = 0.$$

So, if we put $q = \sum_\alpha s_\alpha f(j_\alpha)$, then $cq = 0$. If we choose $L \in \mathcal{L}(\overline{\sigma})$, with $Lq \subseteq S$, then $c(Lq) \subseteq cSq = Scq = 0$. So, $Lq = 0$ and then $q \in \overline{\sigma} Q_{\overline{\sigma}}(S) = 0$. This proves that g is well-defined.

Since $Q_{\overline{\sigma}}(S)$ is $\overline{\sigma}$-closed and g is obviously left S-linear, g extends to some left S-linear morphism $\overline{g} : S \to Q_{\overline{\sigma}}(S)$. Let $\overline{f}$ denote the composition

$$R \hookrightarrow S \xrightarrow{\overline{g}} Q_{\overline{\sigma}}(S).$$

If $j \in I$, then $\overline{f}(j) = \overline{g}(j) = g(j) = f(j)$, hence $\overline{f}$ extends f. Finally, since $\overline{f}$ is obviously left R-linear, this proves that $_R Q_{\overline{\sigma}}(S)$ is σ-closed, indeed. $\square$

Note that the above assumptions are satisfied for any radical σ in $R-\mathbf{mod}$, when R is a prime pi ring. Indeed, in this case any nonzero ideal of R intersects the center $Z(R)$ of R non-trivially, cf. [72].

Let us now specialize to the case where $S = R/K$, for some twosided ideal K of R.

(5.5) Proposition. *Let σ be an arbitrary radical in $R-\mathbf{mod}$, let K be a twosided ideal of R and let $S = R/K$. Then the following assertions are equivalent:*

(5.5.1) $K/KI \in \mathcal{T}_\sigma$ *for all* $I \in \mathcal{L}(\sigma)^1$

(5.5.2) $K \otimes_R M \in \mathcal{T}_\sigma$ *for all* $M \in \mathcal{T}_\sigma$;

(5.5.3) *if* $KM = 0$ *for some left R-module M, then $KQ_\sigma(M) = 0$;*

(5.5.4) *if* $M \in R/K-\mathbf{mod}$, *then* $Q_\sigma(M) \in R/K-\mathbf{mod}$;

(5.5.5) $KQ_\sigma(R) \subseteq Q_\sigma(KI)$ *for all* $I \in \mathcal{L}(\sigma)$.

Moreover, if R is left or right noetherian and if σ is symmetric, then these assertions are equivalent to:

(5.5.6) *for all* $I \in \mathcal{L}(\sigma)$, *there exists some* $J \in \mathcal{L}(\sigma)$ *such that* $JK \subseteq KI$.

Proof. To show that (1) implies (2), consider $k \in K$ and $m \in M$, then there exists some $I \in \mathcal{L}(\sigma)$ with $Im = 0$ and some $J \in \mathcal{L}(\sigma)$ such that $Jk \subseteq KI$. If $j \in J$ and $jk = \sum_\alpha k_\alpha i_\alpha$ for some $k_\alpha \in K$ and $i_\alpha \in I$, then

$$j(k \otimes m) = (\sum_\alpha k_\alpha i_\alpha) \otimes m = \sum_\alpha k_\alpha \otimes i_\alpha m = 0,$$

so $J(k \otimes m) = 0$ and $k \otimes m \in \sigma(K \otimes M)$. This yields that $K \otimes_R M \in \mathcal{T}_\sigma$.

Conversely, if (2) holds and $I \in \mathcal{L}(\sigma)$, then $R/I \in \mathcal{T}_\sigma$, so $K/KI = K \otimes R/I \in \mathcal{T}_\sigma$, indeed. This proves that (2) implies (1).

---

[1]If this holds for *all* twosided ideals K, then σ is *ideal invariant*, in the sense of [69])

Next, assume that (1) holds and let us verify (3). Since $KM = 0$ implies that $K(M/\sigma M) = 0$, we may assume that M is σ-torsionfree. Let $q \in Q_\sigma(M)$, then $Iq \subseteq M$ for some $I \in \mathcal{L}(\sigma)$ and by assumption $K/KI \in \mathcal{T}_\sigma$. Let $k \in K$ and pick $L \in \mathcal{L}(\sigma)$ with $Lk \subseteq KI$, then $Lkq \subseteq KIq \subseteq KM = 0$. So, $kq \in \sigma Q_\sigma(M) = 0$ and since this holds for all $k \in K$, this shows that $Kq = 0$, hence $KQ_\sigma(M) = 0$, indeed.

Conversely, assume that (3) holds. Let $I \in \mathcal{L}(\sigma)$, then it is clear that $Q_\sigma(I/KI) = Q_\sigma(R/KI)$, hence $KQ_\sigma(R/KI) = 0$, since $K(I/KI) = 0$. We thus obtain that $K((R/KI)/\sigma(R/KI)) = 0$. So, $K/KI = K(R/KI) \subseteq \sigma(R/KI)$, showing that $K/KI \in \mathcal{T}_\sigma$, indeed. This proves that (3) implies (1).

The equivalence of (3) and (4) being obvious, let us now show that (1) implies (5). Indeed, pick $I \in \mathcal{L}(\sigma)$, then $K/KI \in \mathcal{T}_\sigma$, by (1), hence $Q_\sigma(K) = Q_\sigma(KI)$. On the other hand, by (3), which is equivalent to (1), it follows from $K(R/K) = 0$ that $KQ_\sigma(R/K) = 0$, hence that $K(Q_\sigma(R)/Q_\sigma(K)) = 0$, as $Q_\sigma(R)/Q_\sigma(K) \subseteq Q_\sigma(R/K)$. But then, $KQ_\sigma(R) \subseteq Q_\sigma(K) = Q_\sigma(KI)$, indeed.

Conversely, assume (5), then $\overline{K} = K/\sigma K \subseteq Q_\sigma(KI)$, for any $I \in \mathcal{L}(\sigma)$, hence $\overline{K}/\overline{KI} \in \mathcal{T}_\sigma$. But then we also have $K/KI \in \mathcal{T}_\sigma$, proving that (5) implies (1).

Finally, assume R to be *left* noetherian and σ to be symmetric. Then $K = \sum_i Rk_i$ for some finite set of $k_i \in K$. If (1) holds and $I \in \mathcal{L}(\sigma)$, then $K/KI \in \mathcal{T}_\sigma$, so we may find some $J \in \mathcal{L}(\sigma)$ such that $Jk_i \in KI$, for all i. But then $JK \subseteq KI$, proving that (1) implies (6) in the left noetherian case.

On the other hand, still assuming σ to be symmetric, but R to be *right* noetherian, then, again, $K = \sum_i k_i R$ for a finite set of $k_i \in K$. If $I \in \mathcal{L}(\sigma)$ and (1) holds, then $K/KI \in \mathcal{T}_\sigma$, so we may find some $J \in \mathcal{L}(\sigma)$ with $Jk_i \subseteq KI$ for all i. But then $JK \subseteq KI$, and since this remains valid for any $I \in \mathcal{L}(\sigma)$, since σ is symmetric, this proves that (1) also implies (6) in the right noetherian case.

Finally, (6) obviously implies (1), so this finishes the proof. $\square$

(5.6) Note. Assuming R to be noetherian, the interested reader may easily verify (with the terminology introduced in the next Chapters), that (5.5.6) implies that σ_K is σ-compatible, (where, for any twosided ideal K of R (as before) σ_K is the radical in $R-\mathbf{mod}$ with Gabriel filter generated by the positive powers K^n of K) and that a symmetric biradical σ in $R-\mathbf{mod}$ satisfies (5.5.6) for any ideal K of R, hence is ideal invariant.

The next result strengthens [40, (3.7)]:

(5.7) Corollary. Let σ be an radical in $R-\mathbf{mod}$, let K be an ideal of R and let $S = R/K$. Then the following assertions are equivalent:

(5.7.1) $KQ_\sigma(R/K) = 0$;

(5.7.2) $Q_\sigma(R/K)$ has a ring structure, making the canonical morphism $R/K \to Q_\sigma(R/K)$ into a ring morphism;

(5.7.3) there exists a canonical isomorphism $Q_\sigma(R/K) = Q_{\bar\sigma}(R/K)$ of left R-modules;

(5.7.4) $Q_\sigma(K) = Ann_{Q_\sigma(R)}(Q_\sigma(R/K))$.

Proof. The equivalence of (1), (2) and (3) follows from (5.3). On the other hand, since we may obviously assume R to be σ-torsionfree, it follows that (4) implies (1), as $K \subseteq Q_\sigma(K)$. Conversely, assume that (2) holds and let us prove (4). Since $Q_\sigma(R) \to Q_\sigma(R/K)$ is then easily verified to be a ring homomorphism, $Q_\sigma(K)$ is a twosided ideal of $Q_\sigma(R)$ and $Q_\sigma(K)Q_\sigma(R/K) = 0$. Moreover, if $q \in Q_\sigma(R)$ and $qQ_\sigma(R/K) = 0$, then $q(Q_\sigma(R)/Q_\sigma(K)) = 0$, as $Q_\sigma(R)/Q_\sigma(K) \subseteq Q_\sigma(R/K)$. So, $qQ_\sigma(R) \subseteq Q_\sigma(K)$, i.e., $q \in Q_\sigma(K)$, which proves the assertion. $\square$

6.   Functorial behaviour

(6.1)   It is clear that if $\varphi : R \to S$ is a centralizing resp. strongly normalizing extension and if σ is a radical resp. a symmetric radical in $R-\mathbf{mod}$ satisfying one of the equivalent conditions of (5.3), then φ extends (uniquely) to a ring homomorphism $\overline{\varphi} = Q_\sigma(\varphi) : Q_\sigma(R) \to Q_{\overline{\sigma}}(S)$, making the following diagram commutative

$$
\begin{array}{ccc}
R & \xrightarrow{\ \varphi\ } & S \\
{\scriptstyle j_\sigma}\downarrow & & \downarrow{\scriptstyle j_{\overline{\sigma}}} \\
Q_\sigma(R) & \xrightarrow[\ \overline{\varphi}\ ]{} & Q_{\overline{\sigma}}(S)
\end{array}
$$

Indeed, up to replacing R by $R/\sigma R$ and S by $S/\overline{\sigma}S = S/\sigma S$, we may clearly assume R and S to be torsionfree. Since $Q_{\overline{\sigma}}(S) = Q_\sigma(S)$ is σ-closed and since $j_\sigma : R \to Q_\sigma(R)$ has σ-torsion cokernel, there exists a unique R-linear map $\overline{\varphi} : Q_\sigma(R) \to Q_{\overline{\sigma}}(S)$ with $\overline{\varphi}j_\sigma = j_{\overline{\sigma}}\varphi$. Let us verify that $\overline{\varphi}$ is a ring homomorphism, to prove the assertion.

Pick $q, q' \in Q_\sigma(R)$ and choose $J \in \mathcal{L}(\sigma)$, in the centralizing case, resp. $J \in \mathcal{L}^2(\sigma)$, in the strongly normalizing case, (where σ has to be symmetric). For any $j \in J$, we then have

$$
\varphi(j)\overline{\varphi}(q)\overline{\varphi}(q') = \overline{\varphi}(jq)\overline{\varphi}(q') = \varphi(jq)\overline{\varphi}(q') =
$$
$$
= \overline{\varphi}((jq)q') = \varphi(j)\overline{\varphi}(qq'),
$$

and we thus obtain that $\varphi(j)(\overline{\varphi}(q)\overline{\varphi}(q') - \overline{\varphi}(qq')) = 0$. Hence

$$
S\varphi(J)(\overline{\varphi}(q)\overline{\varphi}(q') - \overline{\varphi}(qq')) = 0,
$$

so $\overline{\varphi}(q)\overline{\varphi}(q') = \overline{\varphi}(qq')$, since $S\varphi(J) \in \mathcal{L}(\overline{\sigma})$ and since $Q_{\overline{\sigma}}(S)$ is obviously $\overline{\sigma}$-torsionfree.

(6.2)   Of course, the conditions in (5.3) are sufficient to derive the existence of the ring homomorphism $\overline{\varphi} : Q_\sigma(R) \to Q_{\overline{\sigma}}(S)$, but they are obviously not necessary. We will study in this section weaker conditions under which the existence of φ may be proven.

An arbitrary ring homomorphism $\varphi : R \to S$ always factorizes as

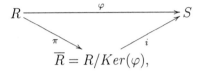

$$\overline{R} = R/Ker(\varphi),$$

where π is surjective and where i is injective and centralizing resp. strongly normalizing, whenever φ is. So, we may clearly separately study the injective and the surjective case, to infer information about general centralizing resp. strongly normalizing ring homomorphisms.

A. The surjective case

Let us start by giving an easy (counter)example, showing that a surjective ring homomorphism $\varphi : R \to S$ does not necessarily induce a ring homomorphism $\overline{\varphi} : Q_\sigma(R) \to Q_{\overline{\sigma}}(S)$, for an arbitrary radical σ in $R-\mathbf{mod}$.

(6.3) Example. Let D be a (commutative) discrete valuation ring with maximal ideal tD. Consider the surjective map

$$\varphi : R = \begin{pmatrix} D\,tD \\ D\ D \end{pmatrix} \to S = R/M = \begin{pmatrix} D\,tD \\ D\ D \end{pmatrix} \Big/ \begin{pmatrix} tD\,tD \\ D\ D \end{pmatrix} = K,$$

where $K = D/tD$ is the residue field of D. Clearly the radical $\sigma = \sigma_{R\backslash M}$ in $R-\mathbf{mod}$ induces the (trivial) radical $\overline{\sigma} = \overline{\sigma_{R\backslash M}} = \sigma_{S\backslash 0}$ in $S-\mathbf{mod}$ and, of course, $Q_{\overline{\sigma}}(S) = S = K$. We claim that there is *no* ring homomorphism

$$\overline{\varphi} : Q_{\sigma_{R\backslash M}}(R) = \begin{pmatrix} D\ tD \\ t^{-1}D\ D \end{pmatrix} \to K = Q_{\overline{\sigma}}(S)$$

extending φ. Indeed, assume that $\overline{\varphi}$ exists and consider the elements

$$u = \begin{pmatrix} 0\,t \\ 0\,0 \end{pmatrix}, v = \begin{pmatrix} 0\ 0 \\ t^{-1}\,0 \end{pmatrix} \in \begin{pmatrix} D\ tD \\ t^{-1}D\ D \end{pmatrix}.$$

Clearly $\overline{\varphi}(u) = \varphi(u) = 0$ and $\overline{\varphi}(v) = 0$, since $\overline{\varphi}(v)^2 = \overline{\varphi}(v^2) = 0$. So, $\overline{\varphi}(u + v) = 0$. However, $(u + v)^2 = 1$, so we should also have $\overline{\varphi}(u + v)^2 = 1$, a contradiction.

On the other hand, as a first positive result, let us mention:

(6.4) Proposition. *Let K be a twosided ideal of the noetherian ring R and let $p : R \to S = R/K$ denote the canonical surjection. If σ is a symmetric ideal-invariant radical in $R-$**mod**, then there exists a unique ring homomorphism $\overline{p} : Q_\sigma(R) \to Q_{\overline{\sigma}}(S)$ extending p, where $\overline{\sigma}$ denotes the radical in $R-$**mod** induced by σ.*

Proof.   It suffices to show that $Q_{\overline{\sigma}}(S) = Q_{\overline{\sigma}}(R/K)$ is σ-closed, as pointed out previously. So, let $I \in \mathcal{L}(\sigma)$ and consider a left R-linear map $f : I \to Q_{\overline{\sigma}}(S)$, which we want to extend to R. Since $Q_{\overline{\sigma}}(S)$ is σ-torsionfree, this will finish the proof. We may obviously assume that $I \in \mathcal{L}^2(\sigma)$, since σ is symmetric. Define $g : \overline{I} = p(I) \to Q_{\overline{\sigma}}(S)$, by letting $g(p(i)) = f(i)$, for any $i \in I$. We have to verify that g is well-defined, i.e., we have to check that if $p(k) = 0$ (or, equivalently, if $k \in K \cap I$), then $f(k) = 0$. Fix $k \in K \cap I$, then, by assumption, we may find $J \in \mathcal{L}^2(\sigma)$ such that $JK \subseteq KI$. So, for any $j \in J$, we have $jk = \sum_\alpha k_\alpha i_\alpha$ for some $k_\alpha \in K$ and $i_\alpha \in I$. But then

$$jf(k) = f(jk) = f(\sum_\alpha k_\alpha i_\alpha) = \sum_\alpha k_\alpha f(i_\alpha) = 0.$$

It follows that $Jf(k) = 0$, i.e., that $f(k) \in \sigma Q_{\overline{\sigma}}(S) = 0$, which proves that g is well-defined, indeed. Now, since g is obviously left S-linear and since $\overline{I} \in \mathcal{L}^2(\overline{\sigma})$, the map g extends to some left S-linear morphism $\overline{g} : S \to Q_{\overline{\sigma}}(S) = 0$. Finally, $\overline{f} = \overline{g}p : R \to Q_{\overline{\sigma}}(S) = 0$ is the desired extension of f, as one easily verifies. $\square$

Let us now prove some general functoriality results in the surjective case. We start with the following straightforward Lemma:

(6.5) Lemma. *Let σ be an radical in $R-\mathbf{mod}$, let K be a twosided ideal of R and let $S = R/K$. Then, for any $M \in S-\mathbf{mod}$ we have*

$$Q_{\overline{\sigma}}(M) = \{q \in Q_\sigma(M); \ Kq = 0\}.$$

Proof. Since $M/\sigma M = M/\overline{\sigma} M$, we may assume M to be torsionfree. Let

$$T = \{q \in Q_\sigma(M); \ Kq = 0\},$$

then clearly T is a left S-module and $M \hookrightarrow T$ is a $\overline{\sigma}$-isomorphism, as $T/M \subseteq Q_\sigma(M)/M \in \mathcal{T}_\sigma$. Finally, T is $\overline{\sigma}$-injective. Indeed, let $L/K \in \mathcal{L}(\overline{\sigma})$, with $L \in \mathcal{L}(\sigma)$ and consider an S-linear map $f : L/K \to T$. We may then find a left R-linear map $f_1 : R \to Q_\sigma(M)$ making the following diagram commute:

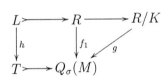

Here $h : L \xrightarrow{p} L/K \xrightarrow{f} T$, where $p : L \to L/K$ is the canonical surjection.

Now, $f_1(K) = fp(K) = 0$, so f_1 factorizes through $g : S = R/K \to Q_\sigma(M)$. Let $x \in S$, then $Kg(x) = g(Kx) = 0$, so g is actually a map $S \to T \subseteq Q_\sigma(M)$, which clearly uniquely extends f. This shows that $T = Q_{\overline{\sigma}}(M)$, indeed.   □

We may now prove:

(6.6) Proposition. *Let K be a twosided ideal of R, let $p : R \to R/K = S$ denote the canonical morphism and let σ be a radical in $R-\mathbf{mod}$, then the following assertions are equivalent:*

(6.6.1) *the canonical morphism*

$$\overline{p} = Q_\sigma(p) : Q_\sigma(R) \to Q_\sigma(R/K) = Q_\sigma(S)$$

factorizes through $Q_{\bar{\sigma}}(S) \subseteq Q_\sigma(S)$, where $\bar{\sigma}$ denotes the induced radical in $S-\mathbf{mod}$;

(6.6.2) $Q_\sigma(K)$ is a twosided ideal of $Q_\sigma(R)$;

(6.6.3) $KQ_\sigma(R) \subseteq Q_\sigma(K)$;

(6.6.4) $Q_\sigma(K) = Ann_{Q_\sigma(R)}(Q_\sigma(R)/Q_\sigma(K))$.

Proof. It is clear that (2), (3) and (4) are equivalent. By the previous Lemma, we know that

$$Q_{\bar{\sigma}}(R/K) = \{q \in Q_\sigma(R/K); \ Kq = 0\}.$$

So, if we assume (1), then $0 = K\bar{p}(Q_\sigma(R)) = \bar{p}(KQ_\sigma(R))$, i.e.,

$$KQ_\sigma(R) \subseteq Ker(\bar{p}) = Q_\sigma(K).$$

Hence (1) implies (3).

Conversely, assume (3) holds, then, from $KQ_\sigma(R) \subseteq Q_\sigma(K)$, it follows that $KIm(\bar{p}) = K(Q_\sigma(R)/Q_\sigma(K)) = 0$. So, $Im(\bar{p}) \subseteq Q_{\bar{\sigma}}(R/K)$, indeed, which shows that (3) implies (1).   □

As a typical example, let us mention:

(6.7) Proposition. [63] Let R be a noetherian prime pi ring and let P be a prime ideal of R with the property that the associated symmetric radical $\sigma = \sigma_{R\setminus P}$ induces a perfect localization. Then the following assertions are equivalent:

(6.7.1) $Q_\sigma(P)$ is a twosided ideal of $Q_\sigma(R)$ (i.e., P satisfies the conditions of the previous result);

(6.7.2) P is left localizable.

Proof. Since R is an FBN ring, it follows from [52, 13.6.6] that σ coincides with σ_C, where $C = \mathcal{C}_R(P)$, the multiplicatively closed subset of R consisting of the elements which are regular modulo P. If $S = Q_\sigma(R)$, then, by [47, 5.6], we know that P is left localizable if and only if $Q = Q_\sigma(P)$ is the Jacobson radical of S and if S/Q is artinian. So, if P is left localizable, then $Q_\sigma(P)$ is certainly a twosided ideal of $Q_\sigma(R)$.

Conversely, let us assume Q to be a twosided ideal of S and let us show Q to be the unique maximal ideal of S. So, suppose M to be a maximal ideal of S, then M is σ-closed, since σ induces a perfect localization, If $M \neq Q$, then $M \not\subseteq Q$ or, equivalently, $M \cap R \not\subseteq P$, i.e., $M \cap R \in \mathcal{L}(\sigma)$. But then $M = Q_\sigma(M \cap R) = S$, a contradiction.

By Posner's Theorem [52, 72], the classical ring of fractions $Q_{cl}(R)$ of R is a pi ring, hence so is S, being a subring of $Q_{cl}(R)$. Essentially by Kaplansky's Theorem, (cf. [52, 12.3.8], for example), it then follows that Q is the Jacobson radical of S and that S/Q is a simple artinian ring, indeed.   □

B. The injective case

(6.8) Let us now specialize to the injective case. We have already seen in (5.4) that if $R \subseteq S$ is a strongly normalizing extension with S and R prime and if σ is a symmetric radical in $R-\mathbf{mod}$ with the property that any $I \in \mathcal{L}(\sigma)$ intersects the center $Z(R)$ of R non-trivially, then $Q_{\bar\sigma}(S)$ is σ-closed. It thus follows that under these assumptions $R \subseteq S$ extends to a ring homomorphism $Q_\sigma(R) \subseteq Q_{\bar\sigma}(S)$.

If we assume R and S to be noetherian, then it appears that we do not have to impose any restriction on the symmetric radical σ.

Let us start with the following Lemma, inspired by a related result in [67] (proved there in the finite case).

(6.9) Lemma. *Let $R \subseteq S$ be a strongly normalizing extension and assume R and S to be prime. If $\mathcal{C}_R$ resp. $\mathcal{C}_S$ denote the Ore sets of regular elements in R resp. S, then $\mathcal{C}_R \subseteq \mathcal{C}_S$.*

Proof. Assume $c \in \mathcal{C}_R$, then we want to show that $c \in \mathcal{C}_S$ as well. So, to prove that c is not a right zero-divisor, for example, suppose that

$sc = 0$ for some

$$s = \sum_{i=1}^{n} r_i z_i = \sum_{i=1}^{n} z_i r'_i \in S,$$

with $0 \neq z_i \in N_R^s(S)$ and $r_i, r'_i \in R$ and let us show that this implies $s = 0$. Choose $1 \leq k \leq n$ such that

$$T = \sum_{i=1}^{k} R z_i = \bigoplus_{i=1}^{k} R z_i = \bigoplus_{i=1}^{k} z_i R$$

as an R-subbimodule of S and such that for any $j > k$ we have $R z_j \cap T \neq 0$. Let $I_j = (T : z_j)$ for any $j > k$, then I_j is a nonzero twosided ideal of R. Denote by J the nonzero twosided ideal $I_{k+1} \cap \ldots \cap I_n$ of R, then $Js \subseteq T$.

On the other hand, if $t = \sum_{i=1}^{k} z_i t_i \in T$ with $t_i \in R$, then $tc = 0$ implies $t = 0$. Indeed, if $tc = \sum_i z_i t_i c = 0$, then, for all i, we have $z_i t_i c = 0$. So, $z_i(R t_i cR) S = R z_i t_i cS = 0$, hence $z_i S(R t_i cR) = 0$, as $R t_i cR$ is a twosided ideal of R and $R \subseteq S$ is strongly normalizing. But then, since S is prime, we obtain that $t_i c = 0$, hence $t_i = 0$, as c is regular. As this holds for all i, this shows that $t = 0$, as claimed.

To conclude, since $Jsc = 0$ and $Js \subseteq T$, it follows that $Js = 0$, hence $JSs = SJs = 0$, which shows that $s = 0$, indeed, since S is prime. This proves the assertion. $\qquad\qquad\qquad\qquad\qquad\qquad\qquad\qquad\qquad\qquad\square$

If we assume R to be left noetherian and prime, then $\mathcal{C}_R$ is an Ore set and R has a classical ring of fractions $Q_{cl}(R) = \mathcal{C}_R^{-1} R$.

It thus follows:

(6.10) Corollary. Let $R \subseteq S$ be a strongly normalizing extension between left noetherian prime rings, then $Q_{cl}(R) \subseteq Q_{cl}(S)$.

We may now prove:

(6.11) Proposition. Let $R \subseteq S$ be a strongly normalizing extension between left noetherian prime rings. If σ is a symmetric radical in R-**mod**, then $Q_\sigma(R) \subseteq Q_{\bar{\sigma}}(S)$.

Proof. Denote by σ_R the radical in R-**mod** associated to the Ore set $\mathcal{C}_R$, i.e., let

$$\sigma_R M = \{m \in M; \; \exists c \in \mathcal{C}_R, \; cm = 0\}$$

for any $M \in R-$**mod**. It is then clear that for any symmetric radical σ in $R-$**mod**, we have $\sigma \leq \sigma_R$. Indeed, $\mathcal{L}(\sigma_R)$ consists of all essential left R-ideals, hence obviously $\mathcal{L}(\sigma) \subseteq \mathcal{L}(\sigma_R)$, as nonzero twosided ideals in a prime ring are clearly essential.

It thus follows that $Q_\sigma(R) \subseteq Q_{cl}(R)$ and $Q_{\bar{\sigma}}(S) \subseteq Q_{cl}(S)$ (as $\bar{\sigma}$ is symmetric as well, by (4.10)). Moreover, by the previous result, $Q_{cl}(R) \subseteq Q_{cl}(S)$.

Now, if $q \in Q_\sigma(R)$, then $q \in Q_{cl}(R) \subseteq Q_{cl}(S)$ and $Iq \subseteq R$ for some $I \in \mathcal{L}^2(\sigma)$. Since $SI \in \mathcal{L}^2(\bar{\sigma})$ (essentially, by the proof of (4.10)) and since $SIq \subseteq S$, we thus obtain that $q \in Q_{\bar{\sigma}}(S)$, indeed. Hence the canonical inclusion $Q_{cl}(R) \subseteq Q_{cl}(S)$ maps $Q_\sigma(R)$ into $Q_{\bar{\sigma}}(S)$, which proves the assertion. □

(6.12) Proposition. *Let $\varphi : R \to S$ be a strongly normalizing extension between prime rings and let σ be a symmetric radical in $R-$**mod** with induced radical $\bar{\sigma}$ in $S-$**mod**.*

Assume one of following conditions holds:

(i) R and S are left noetherian;

(ii) any $0 \neq I \in \mathcal{L}(\sigma)$ intersects the center $Z(R)$ of R non-trivially.

Then the following assertions are equivalent:

(6.12.1) *$\varphi : R \to S$ extends to some ring homomorphism $\bar{\varphi} : Q_\sigma(R) \to Q_{\bar{\sigma}}(S)$ (which is unique as such);*

(6.12.2) *$Q_\sigma(Ker(\varphi)) = Ann_{Q_\sigma(R)}(Q_\sigma(R)/Q_\sigma(Ker(\varphi)))$;*

(6.12.3) *$Q_\sigma(Ker(\varphi))$ is a twosided ideal of $Q_\sigma(R)$.*

Proof. Since $Ker(\varphi) = \varphi^{-1}(0) \in Spec(R)$ as φ is strongly normalizing and S is prime, this follows immediately from (5.4), (6.6), (6.11) and the factorization $\varphi : R \to R/Ker(\varphi) \hookrightarrow S$. □

7. Extensions and localization at prime ideals

(7.1) Although some of the results of this Section (such as (7.11), (7.13), (7.15) and (7.17)) remain valid in a more general context (e.g., only assuming rings to be *left* noetherian), for simplicity's sake, *in this Section, we will assume throughout rings to be noetherian.*

We will mainly be interested in studying the behaviour of classical localization with respect to ring extensions. Since centralizing, normalizing, ... extensions may be defined in terms of centralizing, normalizing, ... bimodules, it should bear no surprise that many of our results will be formulated in terms of bimodules. In particular, one of our main results states that, under suitable (harmless) extra assumptions, the left and right localization at a clique of a strongly normalizing bimodule coincide.

(7.2) Let us start with some preliminaries. Let M be a left R-module and $m \in M$. We call m (left) *weakly regular* if it has the property that for any $r \in R$ and $P \in Spec(R)$, we have $r \in P$ if and only if $rm \in Pm$. We call m (left) *regular* if $rm = 0$ implies $r = 0$, for any $r \in R$, i.e., if $Ann_R^\ell(m) = 0$. Clearly, any regular element $m \in M$ is weakly regular (whence the terminology!). Indeed, if $rm \in Pm$ for some $r \in R$ and some $P \in Spec(R)$, then $rm = pm$ for some $p \in P$, hence $r = p \in P$, as $Ann_R^\ell(m) = 0$.

Note also:

(7.3) Lemma. Let $R \subseteq S$ be a finite strongly normalizing extension. Then any $s \in N_R^s(S)$ is weakly regular.

Proof. If $rs \in Ps$, then $rs = ps$ for some $p \in P$, i.e., $r - p \in Ann_R^\ell(s)$. However, by (3.16), $Ann_R^\ell(s) \subseteq rad(R) \subseteq P$, hence $r - p \in P$ and so $r \in P$, indeed. $\qquad\square$

Of course, it is also obvious that if $R \subseteq S$ is an arbitrary strongly normalizing extension, where S (and hence R) is prime, then any $s \in N_R^s(S)$ is even regular, as pointed out essentially in (3.16).

The need for weakly regular elements is motivated by the following useful result:

(7.4) Proposition. *Let M be an R-bimodule and let $m \in N_R^s(M)$. Assume m to be weakly regular and let r, $r' \in R$ be linked by $mr = r'm$. Then:*

(7.4.1) *for any $P \in Spec(R)$, we have $r \in P$ if and only if $r' \in P$;*

(7.4.2) *for any $X \subseteq Spec(R)$, we have $r \in C_R(X)$ if and only if $r' \in C_R(X)$.*

Proof. To prove the first assertion, note that $r \in P$ implies that

$$r'm = mr \in mP = Pm,$$

hence $r' \in P$. The other implication is similar.

For the second statement, recall that

$$C_R(X) = \bigcap_{P \in X} C_R(P),$$

so it suffices to prove the assertion for $X = \{P\}$.

Assume $r \in C_R(P)$, let $t \in R$ and suppose that $tr' \in P$. Then $tr'm \in Pm = mP$. With $tm = mt'$, we get that

$$mt'r = tmr = tr'm \in mP.$$

It then follows that $t'r \in P$, so $t' \in P$, as r belongs to $C_R(P)$. Hence $t \in P$, by the first part. $\square$

(7.5) Corollary. *Let M be an R-bimodule, which is generated by weakly regular elements in $N_R^s(M)$ – in particular, M will then be a strongly normalizing R-bimodule. Let $X \subseteq Spec(R)$ and assume that*

$C = C_R(X)$ is a left Ore set in R, then:

(7.5.1) for any $m \in M$ and any $c \in C$, there exist $m' \in M$ and $d \in C$, such that $dm = m'c$;

(7.5.2) if M is finitely generated, then for any $m \in M$ and $c \in C$ with $mc = 0$, there exists some $d \in C$ such that $dm = 0$.

Proof. To prove the first assertion, let $c \in C$ and write $m \in M$ as $m = \sum_\alpha r_\alpha m_\alpha$, with $r_\alpha \in R$ and where the $m_\alpha \in N_R^s(M)$ are weakly regular. By the previous result, for any index α, we may find $c'_\alpha \in C$, such that $m_\alpha c = c'_\alpha m_\alpha$. Applying the common denominator property to the r_α and $c'_\alpha \in C$, we may find $d \in C$ and $s_\alpha \in R$, such that $dr_\alpha = s_\alpha c'_\alpha$, for any index α. So,

$$dm = \sum_\alpha dr_\alpha m_\alpha = \sum_\alpha s_\alpha c'_\alpha m_\alpha = \sum_\alpha s_\alpha m_\alpha c = m'c,$$

with $\sum_\alpha s_\alpha m_\alpha = m' \in M$.

To prove the second statement in the finitely generated case, the usual trick applies. Indeed, consider for any positive integer t the left R-submodule

$$M_t = \{ m \in M; \ mc^t = 0 \}$$

of M. Since M is finitely generated, the ascending chain $\{M_t; \ t \in \mathbf{N}\}$ stops, i.e., for some positive integer n, we have $M_n = M_{n+1}$. Applying the first statement then yields the existence of $d \in C$ and $m' \in M$, such that $m'c^n = dm$. But then $m'c^{n+1} = dmc = 0$ (since $mc = 0$), so $m' \in M_{n+1} = M_n$, and $dm = m'c^n = 0$, indeed.   □

We will call (7.5.1) resp. (7.5.2) the *left Ore* resp. the *left reversibility* conditions on C with respect to M. If both are satisfied, then we say that C is a *set of left denominators* for M.

Since any element in $N_R^s(S)$ is weakly regular whenever $R \subseteq S$ is a finite strongly normalizing extension, the previous result yields:

(7.6) Corollary. *Let $R \subseteq S$ be a finite strongly normalizing extension and let $X \subseteq Spec(R)$. If $\mathcal{C}_R(X)$ is a left Ore set in R, then $\mathcal{C}_R(X)$ is also a left Ore set in S.*

Of course, one may sometimes eliminate the finiteness condition in the previous result. Indeed, as we have seen in (6.9), for example, if $R \subseteq S$ is a strongly normalizing extension and if R and S are prime, then $\mathcal{C}_R \subseteq \mathcal{C}_S$.

Note also, that if M is a centralizing R-bimodule, then we do not need any extra regularity assumptions:

(7.7) Lemma. *Let M be a centralizing resp. a finitely generated centralizing R-bimodule and C a left Ore set in R, then C satisfies the left Ore condition resp. the left reversibility condition with respect to M.*

Proof. To prove the left Ore condition, let $m \in M$, then there exist $m_i \in M^R$ and $r_i \in R$ such that

$$m = \sum_i m_i r_i = \sum_i r_i m_i.$$

The usual method then shows that there exist $s_i \in R$ and $d \in C$, such that $dr_i = s_i c$ for all i. So, if we put $m' = \sum_i m_i s_i$, then $dm = m'c$, indeed.

The left reversibility condition follows again from the left Ore condition, since the hypotheses imply M to be noetherian. $\qquad \square$

(7.8) For any subset $X \subseteq Spec(R)$, we define a radical σ_X in $R-\mathbf{mod}$, by letting $\sigma_X = \bigwedge_{P \in X} \sigma_P$, where, as pointed out in Chapter I, for each $P \in Spec(R)$, we denote by σ_P the radical in $R-\mathbf{mod}$ associated to the multiplicatively closed subset $\mathcal{C}_R(P)$ or, equivalently, σ_P is the Lambek-Michler radical associated to P, cf. [47].

The same construction allows us to introduce the analogous radicals ρ_P resp. ρ_X in $\mathbf{mod}-R$.

Denote by Q_X^ℓ resp. Q_X^r the localization functors in $R-\mathbf{mod}$ resp. $\mathbf{mod}-R$ with respect to σ_X resp. ρ_X.

(7.9) Theorem. *Let R be a (noetherian) ring in which every clique is classically localizable (i.e., a noetherian ring satisfying the strong second layer condition and in which every clique satisfies the weak intersection property) and let X be a link closed subset of $Spec(R)$. If M is an R-bimodule, which is either centralizing or generated by weakly regular elements in $N_R^s(M)$, then $Q_X^\ell(M) = Q_X^r(M)$.*

Proof. Since X is link-closed, we may write it as a union of cliques $X = \bigcup_{\alpha \in A} X_\alpha$. In particular, since each X_α satisfies the weak intersection property, it easily follows that

$$\sigma_X = \bigwedge_{P \in X} \sigma_P = \bigwedge_{\alpha \in A} (\bigwedge_{P \in X_\alpha} \sigma_P) = \bigwedge_{\alpha \in A} \sigma_{C_R(X_\alpha)},$$

and similarly for ρ_X. Since localization commutes with direct unions, R being noetherian, we may assume M to be finitely generated, i.e., $M = \sum_i m_i R = \sum_i R m_i$ for a *finite* number of $m_i \in N_R^s(M)$. So, M thus satisfies the left and right reversibility condition (7.5.2) with respect to each X_α, both in the centralizing and the strongly normalizing case. It follows, for each $\alpha \in A$, that $\sigma_{X_\alpha} M = \rho_{X_\alpha} M$, hence that $\sigma_X M = \rho_X M$. In particular, cancelling out the torsion, we may assume, without loss of generality, M to be torsionfree with respect to X, on the left and on the right.

Write $Q = Q_X^\ell(M)$ and fix an index $\alpha \in A$. If $q \in Q$, then we may find some $c \in C(X_\alpha)$, such that $cq = m \in M$. We may also pick $m' \in M$ and $d \in C(X_\alpha)$ such that $md = cm'$. So, $cqd = md = cm'$, hence $cq_1 = 0$, with $q_1 = qd - m' \in Q$.

On the other hand, since Q/M is σ_X-torsion, we may pick $I \in \mathcal{L}(\sigma_X)$ with $Iq_1 \subseteq M$. It is easy to see, using the noetherian hypothesis and the common denominator property, that we may as well assume I to be cyclic, i.e., $I = Ry$, for some $y \in R$, hence $x = yq_1 \in M$. Pick $y' \in R$

and $c' \in \mathcal{C}(X_\alpha)$, such that $c'y = y'c$, then

$$c'x = c'yq_1 = y'cq_1 = 0.$$

So, we may find some $c'' \in \mathcal{C}(X_\alpha)$, with $yq_1c'' = xc'' = 0$. But then $Iq_1c'' = 0$, hence $q_1c'' \in \sigma_X M$. So, $qdc'' = m'c'' \in M$, with $dc'' \in \mathcal{C}(X_\alpha)$. Since $q \in Q$ is arbitrary, this proves that the *right* R-module Q/M is torsion at each X_α, hence that Q/M is ρ_X-torsion.

Finally, note that Q is essential over M as a *right* R-module. Indeed, if $0 \neq q \in Q$, then we may find some $J \in \mathcal{L}(\rho_X)$ with the property that $qJ \subseteq M$. If $qJ = 0$, then, with $I \in \mathcal{L}(\sigma_X)$ such that $Iq \subseteq M$, we obtain that $IqJ = 0$, so $Iq \subseteq \rho_X M = 0$, and $q \in \sigma_X Q = 0$, a contradiction. It follows that $Q = Q_X^\ell(M) \subseteq Q_X^r(M)$, whence equality, by symmetry.   □

(7.10) Corollary. *Let R be a (noetherian) ring in which every clique is classically localizable and let $\varphi : R \to S$ be a ring homomorphism. Assume one of the following properties is satisfied:*

(7.10.1) *φ is a centralizing extension;*

(7.10.2) *φ is a finite strongly normalizing extension;*

(7.10.3) *φ is a strongly normalizing extension and S is prime.*

If $X \subseteq Spec(R)$ is link closed, then the rings $Q_X^\ell(S)$ and $Q_X^r(S)$ are canonically isomorphic.

Proof. In the centralizing case, this follows trivially from the previous result. In the finite strongly normalizing case, factorize the map $\varphi : R \to S$ into the surjection $R \to R/Ker(\varphi)$ (which is centralizing) and the finite normalizing extension $R/Ker(\varphi) \hookrightarrow S$, and apply (7.3) and the previous result. In the prime strongly normalizing case, work as in the finite case, but apply the remarks after (7.3), i.e., use the fact that $N_R^s(S)$ consists of regular elements, when $R \to S$ is strongly normalizing and S is prime.   □

The following result will be needed below:

(7.11) Proposition. Let $R \subseteq S$ be a finite strongly normalizing extension and let $X \subseteq Spec(R)$. Then, with $C = C_R(X)$:

(7.11.1) the induced ring homomorphism $C^{-1}R \subseteq C^{-1}S$ is a strongly normalizing extension;

(7.11.2) for any $Q \in Spec(S)$, we have $C^{-1}(Q \cap R) = C^{-1}Q \cap C^{-1}R$.

Proof. To prove the first assertion, assume that $S = \sum_{i=1}^{n} Ra_i$ for a family $\{a_i;\ 1 \le i \le n\}$ of strongly R-normalizing elements. First note that every twosided ideal J of $C^{-1}R$ is of the form $J = C^{-1}I$ for some twosided ideal I of R. So, for any $i = 1, \ldots, n$, we have

$$Ja_i = C^{-1}Ia_i = C^{-1}a_iI \overset{*}{=} a_iC^{-1}I = a_iJ,$$

where the equality (*) follows from (7.3) and (7.4). Let us now show that $C^{-1}S = \sum_i (C^{-1}R)a_i = \sum_i a_i(C^{-1}R)$. The inclusion $C^{-1}S \subseteq \sum a_i(C^{-1}R)$ is obvious. To prove the other inclusion, consider an arbitrary element $x \in \sum a_i(C^{-1}R)$. We may write x as

$$x = a_1(c_1^{-1}r_1) + \ldots + a_n(c_n^{-1}r_n).$$

As C is a left Ore set in S, there exist some $d \in C$ and $t_1, \ldots, t_n \in S$, such that $da_i = t_ic_i$, for any $i = 1, \ldots, n$. So, $dx = t_1r_1 + \ldots + t_nr_n = t \in S$, hence $x = d^{-1}t \in C^{-1}S$.

For the second statement, note that the inclusion $C^{-1}(Q \cap R) \subseteq C^{-1}Q \cap C^{-1}R$ is obvious. Conversely, let x be an arbitrary element of $C^{-1}Q \cap C^{-1}R$, then $x = c_1^{-1}q = c_2^{-1}r$ for certain $q \in Q$, $r \in R$ and $c_1, c_2 \in C$. Reducing to a common denominator, there exist $r_1, r_2 \in R$ and $c \in C$ with the property that $c_1^{-1} = c^{-1}r_1$ resp. $c_2^{-1} = c^{-1}r_2$, and then $c^{-1}r_1q = c^{-1}r_2r$. So, there exists $d \in C$ such that $dr_1q = dr_2r$. Hence $dr_1q \in Q \cap R$, so $r_1q \in C^{-1}(Q \cap R)$ and finally $x = c_1^{-1}q = c^{-1}r_1q \in C^{-1}(Q \cap R)$, indeed. $\square$

(7.12) Consider a finite strongly normalizing extension $R \subseteq S$. We want to associate to a localizable subset X of prime ideals of R a localizable subset X^* of prime ideals of S.

In order to realize this, associate to any $X \subseteq Spec(R)$ the set $X^* \subseteq Spec(S)$ defined by:

$$X^* = \{Q \in Spec(S); \ Q \cap R \in X\}.$$

One may then prove:

(7.13) Proposition. *Let $R \subseteq S$ be a finite strongly normalizing extension and assume $X \subseteq Spec(R)$ to be localizable. Then $X^* \subseteq Spec(S)$ is localizable and the associated localizations induce a strongly normalizing extension $R_X \subseteq S_{X^*}$.*

Proof. By the previous result, $C = \mathcal{C}_R(X)$ is a left Ore set in S, so we may consider the localization $C^{-1}S$. Let us show that $C^{-1}S$ is the localization of S with respect to X^*.

First, note that for any $Q \in X^*$, we have $C \cap Q = C \cap (Q \cap R) = \varnothing$, since $Q \cap R \in X$.

Next, as $C^{-1}(Q \cap R) = C^{-1}Q \cap C^{-1}R$, we may consider, by passing over to quotients, the finite strongly normalizing extension $C^{-1}R/C^{-1}(Q \cap R) \subseteq C^{-1}S/C^{-1}Q$ and as $C^{-1}R/C^{-1}(Q \cap R)$ is artinian, clearly so is $C^{-1}S/C^{-1}Q$.

Finally, if T is a primitive ideal of $C^{-1}S$, obviously $T = C^{-1}Q$ for some $Q \in Spec(S)$. It follows that $T \cap C^{-1}R = C^{-1}(Q \cap R)$ is primitive in $C^{-1}R$. Hence $(Q \cap R) \in X$, so $Q \in X^*$. $\qquad\square$

In order to generalize the previous result to classically localizable sets, we need a change of rings result for the second layer condition.

The fundamental result below provides a partial answer to this question:

(7.14) Theorem. [49] *If $R \subseteq S$ is a finite extension and if R satisfies the second layer condition, then so does S.*

The previous result is not directly useful to us, if we want to extend (7.13). This is due to the fact that (i) it uses the second layer condition on *both* sides and (ii) it needs the second layer condition on *all* prime ideals of R and not merely on those belonging to a particular subset $X \subseteq Spec(R)$.

However, under the more restrictive hypothesis that the extension be finite and (strongly) normalizing, a straightforward adaptation of its proof, combined with [67, Théorème 8] and [79, Proposition 16] yields:

(7.15) Proposition. *Let $i : R \hookrightarrow S$ be a finite normalizing extension. Let $Q \in Spec(S)$ and let $(^a i)(Q)$ denote the set of prime ideals of R, which are minimal over $Q \cap R$. If*

$$\bigcup \{cl^\ell(P); \ P \in (^a i)(Q)\}$$

satisfies the strong left second layer condition in R, then Q satisfies the strong left second layer condition in S.

Proof.   Assume that Q does not verify the strong left second layer condition, then there exists a short exact sequence of finitely generated uniform left S-modules

$$0 \to L \to M \to N \to 0,$$

with the property that $L = Ann^r_M(Q)$ is torsionfree as a left S/Q-module and that $Ann^\ell_R(M) = Ann^\ell_R(N) = T \in Spec(S)$ with $T \subset Q$. Passing over to quotients, we may assume that $T = 0$, i.e., that S is prime and hence that R is semiprime. Applying [67, théorème 8], we obtain that $\mathcal{C}_R(Q \cap R) \subseteq \mathcal{C}_S(Q)$, so L is torsionfree as a left $R/Q \cap R$-module. On the other hand, applying [45, 4.2.6], we find $ass_S(L) = Q$, so $Ass_R(L) \subseteq (^a i)(Q)$, by [79, Proposition 16].

Let us now consider a left R-submodule K of M, which is maximal with respect to the property $L \cap K = 0$. Then L is isomorphic to an essential left R-submodule of M/K, hence $Ass_R(M/K) = Ass_R(L) \subseteq (^a i)(Q)$. Since, by assumption, $\bigcup \{cl^l(P); \ P \in (^a i)(Q)\}$ satisfies the left second layer condition, we may apply (6.5) in Chapter I to M/K. So, there exist

$$P_1, \ldots, P_n \in \bigcup \{cl^\ell(P); \ P \in (^a i)(Q)\},$$

with the property that $P_1 \ldots P_n(M/K) = 0$.

From here on, the same reasoning as in [49, 2.5] applies. Indeed, since S is prime, Q contains a regular element of S, so Q is essential in S as a left R-module. For, if L is a left R-submodule of S and $c \in Q$ is regular, then it follows from $L \cap Q = 0$ that $cL \subseteq QL \subseteq L \cap Q = 0$, hence $L = 0$. But then $Q \cap R$ is essential in R and as R is semiprime, it follows from (1.6) in Chapter I, that $Q \cap R$ contains a regular element, i.e., $(Q \cap R) \cap C_R \neq \varnothing$. This implies that $P \cap C_R \neq \varnothing$, for any $P \in (^a i)(Q)$. As every P_i belongs to the left clique of some $P \in (^a i)(Q)$, every P_i contains a regular element of R as well.

So, $J = Ann_R^\ell(M/K)$ contains a regular element. Hence S/JS is torsion as a right S-module, since $J \cap C_R \neq \varnothing$ implies that $J \in \mathcal{L}(\rho_{C_R})$, hence that $JS \in \mathcal{L}(\rho_{C_S})$, where ρ_{C_R} resp. ρ_{C_S} are associated to C_R resp. C_S in $\mathbf{mod}{-}R$ resp. $\mathbf{mod}{-}S$. Moreover, since S/JS is left finitely generated, we obtain that its annihilator $I = Ann_S^r(S/JS)$ belongs to the filter $\mathcal{L}(\rho_{C_S})$ and, in particular, does not vanish. Finally, $IM = SIM \subseteq JSM = JM \subseteq K$, so IM is a left S-submodule of M with the property that $IM \cap L = 0$. As M is a uniform left S-module, it follows that $IM = 0$, contradicting $Ann_S^\ell(M) = T = 0$. □

In a similar way:

(7.16) Proposition. Let $R \subseteq S$ be a finite strongly normalizing extension. Let $Q \in Spec(S)$, let $P = Q \cap R$ and assume that $cl^\ell(P)$ satisfies the left second layer condition in R. Then Q satisfies the left second layer condition in S.

Proof. The main difference between the proof of the present result and the previous one resides in the fact that we now have to use (6.7) instead of (6.5) in Chapter I.

Only assuming $R \hookrightarrow S$ to be a finite normalizing extension does not seem to suffice in order to infer the desired result, since at some point we need that $Ann_R^r(L)$ is torsionfree as a left R/P-module for any $P \in ({}^a i)(Q)$ and we only know L is torsionfree over $R/Q \cap R$ – notations are as in the previous proof.

However, if $R \hookrightarrow S$ is *strongly* normalizing, things work out nicely. Indeed, in this case $Q \cap R = P$ is prime and as $\mathcal{C}_R(P) \subseteq \mathcal{C}_S(Q)$, it follows that L is torsionfree as a left R/P-module. Again it follows that $ass_S(L) = Q$, hence $Ass_R(L) = \{P\}$, by [79]. Pick $K \subseteq M$ as in the previous proof, then we find $Ass_R(M/K) = Ass_R(L) = \{P\}$. Moreover, since by assumption $cl^\ell(P)$ satisfies the left second layer condition and since M/K is torsionfree as a left R/P-module (as an essential extension of L), we may now apply (6.7) in Chapter I. So, there exist $P_1, \ldots, P_n \in cl^\ell(P)$, with $P_1 \ldots P_n(M/K) = 0$.

The rest of the proof may be given along the lines of the previous one and even simplifies somewhat, if one uses the fact that R is now prime and that $({}^a i)(Q) = \{P\}$.   □

We are now finally able to prove the result announced previously:

(7.17) Proposition. *Let $R \subseteq S$ be a finite strongly normalizing extension and assume $X \subseteq Spec(R)$ to be classically left localizable. Then $X^* \subseteq Spec(S)$ is classically localizable and the localizations induce a strongly normalizing extension $R_X \subseteq S_{X^*}$.*

Proof. If $X \subseteq Spec(R)$ is classically localizable, then, in particular, X is left link closed and satisfies the left second layer condition. From the previous result, it then follows that X^* satisfies the left second layer condition as well, and as X^* is left localizable by (7.13), this proves that X^* is classically localizable, indeed.

The second assertion has already been proved in (7.13).   □

To conclude this Section, let us now briefly consider the behaviour of links with respect to extensions. Let us first mention:

(7.18) Lemma. *Let R be a left noetherian ring, and $P, Q \in Spec(R)$. If I is a twosided ideal of R with the property that $I \subseteq P \cap Q$, then:*

(7.18.1) *if $Q/I \rightsquigarrow P/I$ (defined by J/I) for some twosided ideal J of R, then $Q \rightsquigarrow P$ (defined by J);*

(7.18.2) *$Q/I \rightsquigarrow P/I$ (defined by 0) if and only if $Q \rightsquigarrow P$ (defined by I).*

Proof. To prove the first assertion, first note that by definition

$$(QP + I)/I = (Q/I)(P/I) \subseteq J/I \subset (Q/I) \cap (P/I),$$

so $QP \subseteq J \subset Q \cap P$. Moreover, as

$$Ann_R^{\ell}((Q \cap P)/I)/(J/I)) = Q/I$$

resp.

$$Ann_R^{r}((Q \cap P)/I)/(J/I)) = P/I,$$

and since $(Q \cap P/I)/(J/I) = Q \cap P/I$ (up to canonical isomorphism), we obtain that $Ann_R^{\ell}(Q \cap P/J) = Q$ resp. $Ann_R^{r}(Q \cap P/J) = P$, and that $Q \cap P/J$ is torsionfree as a left R/Q-module and as a right R/P-module.

One of the implications of the second assertion follows from the first one. On the other hand, if $Q \rightsquigarrow P$ (defined by I), then $QP \subseteq I \subset Q \cap P$, so

$$(Q/I)/(P/I) = (QP + I)/I = 0 \subset Q \cap P/I = (Q/I) \cap (P/I).$$

The other properties proving that $Q/I \rightsquigarrow P/I$ (defined by 0) follow in a similar way. $\square$

The following result is inspired by a similar result, due to [15].

(7.19) Proposition. *Let $R \subseteq S$ be a finite strongly normalizing extension of rings satisfying the second layer condition and consider a pair of prime ideals $Q_1, Q_2 \in Spec(S)$ with the property that $Q_1 \rightsquigarrow Q_2$. Then either $Q_1 \cap R = Q_2 \cap R$ or $Q_1 \cap R \rightsquigarrow Q_2 \cap R$.*

Proof. Assume that $Q_1 \rightsquigarrow Q_2$ via A, then $Q_1/A \rightsquigarrow Q_2/A$ via 0 in R/A, and passing over to quotients, we may assume that $A = 0$, without loss of generality. We thus obtain that $Q_1 \cap Q_2$ is a $(S/Q_1, S/Q_2)$-bimodule, which is torsionfree on both sides. Since $\mathcal{C}_R(Q_1 \cap R) \subseteq \mathcal{C}_S(Q_1)$ and similarly for Q_2, by [67, théorème 8], it follows that $Q_1 \cap Q_2$ is an $(R/Q_1 \cap R, R/Q_2 \cap R)$-bimodule, which is torsionfree and noetherian, on both sides. From [45], it follows that $Kdim(R/Q_1 \cap R) = Kdim(R/Q_2 \cap R)$, where $Kdim$ denotes the classical Krull dimension. So, either $Q_1 \cap R = Q_2 \cap R$, or they are incomparable. If $Q_1 \cap R \neq Q_2 \cap R$, then $Q_1 \cap R$ and $Q_2 \cap R$ are minimal primes of R, since $(Q_1 \cap R)(Q_2 \cap R) = 0$. Moreover, since for any other prime ideal T of R we have $Kdim(R/T) < Kdim(R/Q_i \cap R)$, it follows that the set $\{Q_1 \cap R, Q_2 \cap R\}$ is link closed. Assume there does not exist a link $Q_1 \cap R \rightsquigarrow Q_2 \cap R$, then $\{Q_2 \cap R\}$ will be left link closed and hence classically localizable. From (7.17), it will then follow that

$$\{Q_2 \cap R\}^* = \{P \in Spec(R); \ P \cap R = Q_2 \cap R\}$$

is classically left localizable in S and, in particular, left link closed. As this contradicts the existence of a link $Q_1 \rightsquigarrow Q_2$, together with $Q_1 \notin \{Q_2 \cap R\}^*$, $Q_2 \in \{Q_2 \cap R\}^*$, this proves the assertion $\qquad \square$

The previous result does not generalize to finite normalizing extensions, as shown by an example in [49].

Let us mention as an immediate corollary:

(7.20) Corollary. *Let $R \subseteq S$ be a finite strongly normalizing extension between rings satisfying the second layer condition, then:*

(7.20.1) *if $X \subseteq Spec(R)$ is left link closed, then so is X^*;*

(7.20.2) *for any $P \in Spec(S)$, we have $cl^\ell(P) \subseteq cl^\ell(P \cap R)^*$. In particular, if $cl^\ell(P \cap R)$ is finite, then so $cl^\ell(P)$.*

Proof. To prove the first statement, consider $P \in X^*$ and assume that $P \rightsquigarrow Q$, then either $\dot{Q} \cap R = P \cap R$ or $P \cap R \rightsquigarrow Q \cap R$. In both cases, we obtain $Q \cap R \in X$, so $Q \in X^*$, indeed.

For the second statement, note that $cl^{\ell}(P \cap R)^*$ contains P and is link closed, by the first part. Hence $cl^{\ell}(P \cap R)$ contains $cl^{\ell}(P)$. The final assertion follows from (3.15).   □

As far as lifting links is concerned, let us mention the following results:

(7.21) Proposition. [49] *Let $R \subseteq S$ be a finite extension of rings and let P_1 and P_2 be prime ideals of R with the property that $P_1 \rightsquigarrow P_2$. Then there exist Q_1 resp. Q_2 in $Spec(S)$ lying over P_1 resp. P_2 with the property that either $Q_1 = Q_2$ or that there exists a chain $Q_1 = T_1, T_2, \ldots, T_t = Q_2$ in $Spec(S)$ with $t \geq 2$ and such that $T_i \rightsquigarrow T_{i+1}$ for any $1 \leq i \leq t - 1$.*

(7.22) Corollary. *Let $R \subseteq S$ be a finite strongly normalizing extension and let $X \subseteq Spec(R)$. If X^* is left link closed, then so is X.*

Proof. Given $Q \in Spec(R)$ and $P \in X$ with $P \rightsquigarrow Q$, then the previous result implies the existence of $\overline{Q}$ resp. $\overline{P} \in Spec(S)$, with $\overline{Q} \cap R = Q$ resp. $\overline{P} \cap R = P$, and a chain of prime ideals $\overline{P} \rightsquigarrow T_2 \rightsquigarrow \ldots \rightsquigarrow T_{t-1} \rightsquigarrow \overline{Q}$. Since $P \in X$, we obtain that $\overline{P} \in X^*$. As X^* is left link closed, $\overline{Q} \in X^*$, so $Q \in X$, indeed.   □

Chapter III

STABILITY

> *The reason why noetherian rings behave so nicely within the field of commutative algebra is mainly due to the validity of the Artin-Rees Lemma.*

Of course, this sounds like just another silly aphorism and, of course, many people will strongly disagree with it. But, as we all know, there *is* some truth in this statement!

Now, somewhat painfully for noncommutative ring theory, the Artin-Rees Lemma unfortunately fails to hold for general, noncommutative noetherian rings. On the other hand, although not all twosided ideals of an arbitrary noetherian ring possess the Artin-Rees property, it appears that many twosided ideals of most noetherian rings occurring in nature do – we leave it to the creativity of the reader to give a more precise meaning to these rather fuzzy [97] terms *many* and *most* (some clearer indications about this will, of course, be given below).

In this Chapter, we will start by giving some torsion theoretic interpretation of different versions of the Artin-Rees property and show how they interact modulo some rather harmless restrictions on the rings over which they are defined. In Chapter IV, we also relate these notions to so-called compatibility properties between radicals and the localization functors associated to them. These results yield the framework necessary to develop the construction techniques of structure sheaves, which are the main topic to be considered in the next Chapter.

1.   Stable torsion theories

In this Chapter, unless explicitly indicated otherwise, R will denote an arbitrary, not necessarily noetherian ring.

As before, let us call a radical σ in $R-\mathbf{mod}$ *stable* if the class $\mathcal{T}_\sigma$ is closed under taking injective hulls. There are several alternative ways of describing stable radicals, the most obvious ones being through the following result:

(1.1) Proposition. *Let σ be a radical in $R-\mathbf{mod}$. The following statements are equivalent:*

(1.1.1) *σ is stable;*

(1.1.2) *σM is a direct summand of M, for any injective left R-module M;*

(1.1.3) *σM is a direct summand of M, for any σ-injective left R-module M.*

Proof. Let us first show that (1) implies (2). Let M be an injective left R-module, then $\sigma M \subseteq E(\sigma M) \subseteq M$. By assumption, $E(\sigma M)$ is σ-torsion, so we obtain $E(\sigma M) = \sigma M$. Hence σM is injective and it is a direct summand of M.

Next, assume (2) and let us verify (3). Let M be a σ-injective left R-module. The canonical inclusion $i : \sigma M \hookrightarrow M$ extends to some left R-linear map $f : \sigma E(M) \to M$, as $\sigma E(M)/\sigma M$ is obviously σ-torsion. Now, f is injective, since $0 \neq Ker(f) \subseteq \sigma E(M)$ would imply $0 \neq Ker(f) \cap M \subseteq \sigma M$, hence $\sigma E(M) = \sigma M$. As $E(M)$ is injective, (2) implies that $\sigma M = \sigma E(M)$ is a direct summand of $E(M)$, hence that it is injective. So, σM is a direct summand of M, indeed.

Finally, let us prove that (3) implies (1). Let M be a σ-torsion left R-module, then $\sigma E(M)$ is a direct summand of $E(M)$ containing M. So, since M is essential in $E(M)$, we have $\sigma E(M) = E(M)$. □

Recall that a submodule N of a left R-module M is *essentially closed* or a *complement* in M, if it has no non-trivial essential extensions in M. Equivalently, if it is the trace in M of a direct summand of $E(M)$. With this terminology, it follows from the previous result:

(1.2) Corollary. *A radical σ in $R-$mod is stable if and only if σM is essentially closed in M, for any left R-module M.*

(1.3) Let σ a radical in $R-$mod and M a left R-module. Then

$$\mathcal{L}_\sigma(M) = \{L \leq M; \ M/L \in \mathcal{T}_\sigma\}$$

is a filter of left R-submodules of M, which defines a linear topology on M, by considering $\mathcal{L}_\sigma(M)$ as a neighbourhood basis of $0 \in M$. In particular, $\mathcal{L}_\sigma(R) = \mathcal{L}$.

Let N be a left R-submodule of M. Then the topology defined by $\mathcal{L}_\sigma(M)$ on M induces a topology on N. On the other hand, N is also endowed with its own topology defined by $\mathcal{L}_\sigma(N)$.

These topologies do not coincide, in general. However, they do in the stable case and even provide an alternative way of characterizing stable radicals.

(1.4) Lemma. *Let σ be a radical in $R-$mod. The following statements are equivalent:*

(1.4.1) *σ is stable;*

(1.4.2) *for any left R-module M and any submodule N of M, we have*

$$\mathcal{L}_\sigma(N) = \{M' \cap N; \ M' \in \mathcal{L}_\sigma(M)\}.$$

Proof. Let us first prove that (1) implies (2). So, let σ be stable and consider a pair of left R-modules $N \subseteq M$. Clearly, for any $M' \in \mathcal{L}_\sigma(M)$, we have $M' \cap N \in \mathcal{L}_\sigma(N)$. Conversely, let $N' \in \mathcal{L}_\sigma(N)$ and consider the set

$$\mathcal{S} = \{M' \subseteq M; \ M' \cap N = N'\}.$$

Clearly $\mathcal{S}$ is inductive and non-empty, as $N' \in \mathcal{S}$. Pick a maximal element $M' \in \mathcal{S}$, then

$$N/N' = N/M' \cap N = N + M'/M'.$$

To conclude, we want to verify that $M' \in \mathcal{L}_\sigma(M)$ or, equivalently, that $M/M' \in \mathcal{T}_\sigma$. Now, the maximality of M' in $\mathcal{S}$ is easily seen to imply the extension $N + M'/M' \subseteq M/M'$ to be essential. As N/N' is σ-torsion and as σ is assumed to be stable, it thus follows that M/M' is σ-torsion, indeed.

Conversely, assume (2) and let us prove (1). Let $M \in \mathcal{T}_\sigma$ and consider the extension $M \subseteq E(M)$, where $E(M)$ denotes an injective hull of M. As $0 \in \mathcal{L}_\sigma$, there exists some left R-submodule $L \in \mathcal{L}_\sigma(E(M))$, with $L \cap M = 0$. As the extension $M \subseteq E(M)$ is essential, it follows that necessarily $L = 0$, however, so $E(M) \in \mathcal{T}_\sigma$, indeed. $\square$

(1.5) Proposition. *Let σ be a radical in $R-$**mod**. If σ is stable, then for any indecomposable injective left R-module M either M is σ-torsion or M is σ-torsionfree.*

The converse is also true, if R is left noetherian.

Proof. Let σ be a stable radical in R-**mod** and M an indecomposable injective left R-module. By (1.1), it follows that σM is a direct summand of M, so either $\sigma M = 0$ or $\sigma M = M$.

On the other hand, let us assume now that R is left noetherian and that every indecomposable injective left R-module is either σ-torsion or σ-torsionfree. Let $M \in \mathcal{T}_\sigma$, then, since R is left noetherian, $E(M) = \bigoplus_{\alpha \in A} E_\alpha$, where E_α is an indecomposable injective left R-module for each $\alpha \in A$. Since M is essential in $E(M)$, it follows that $M \cap E_\alpha \neq 0$, hence $E_\alpha \notin \mathcal{F}_\sigma$ for any $\alpha \in A$. Thus, by hypothesis, $E_\alpha \in \mathcal{T}_\sigma$ for all $\alpha \in A$ and $E(M)$ is σ-torsion. □

(1.6) Proposition. *Assume $\{\sigma_\alpha; \ \alpha \in A\}$ is a family of stable radicals in R-mod. Then:*

(1.6.1) $\bigwedge_{\alpha \in A} \sigma_\alpha$ *is stable;*

(1.6.2) *if the ring R is left noetherian, then $\bigvee_{\alpha \in A} \sigma_\alpha$ is stable.*

Proof. To prove the first statement, consider a $\bigwedge_\alpha \sigma_\alpha$-torsion left R-module M. Then M is σ_α-torsion for any $\alpha \in A$, and by the stability of these, $E(M)$ is σ_α-torsion for all $\alpha \in A$. So, $E(M)$ is $\bigwedge_\alpha \sigma_\alpha$-torsion, hence $\bigwedge_\alpha \sigma_\alpha$ is stable, indeed.

On the other hand, if R is a left noetherian ring and if E is an indecomposable injective left R-module, which is not $\bigvee_\alpha \sigma_\alpha$-torsionfree, then E is not σ_α-torsionfree for some $\alpha \in A$. Since σ_α is stable, it follows that E is σ_α-torsion and so it is $\bigvee_\alpha \sigma_\alpha$-torsion as well. This shows that $\bigvee_\alpha \sigma_\alpha$ is stable, which finishes the proof. □

Note that if R is not left noetherian, then the second statement in the previous Proposition fails to be true in general.

When σ is a stable radical, the calculation of the localization is somewhat easier than in the general case:

(1.7) Proposition. *Let σ be a stable radical in R-mod, then for any left R-module we have*

$$Q_\sigma(M) \cong \varinjlim_{I \in \mathcal{L}(\sigma)} Hom_R(I, M).$$

Proof. For any $N \in R$-mod, we put

$$Q_{(\sigma)}(M) = \varinjlim_{I \in \mathcal{L}(\sigma)} Hom_R(I, M).$$

Since direct limits are exact, the exact sequence

$$0 \to \sigma M \to M \to M/\sigma M \to 0$$

induces an exact sequence

$$0 \to Q_{(\sigma)}(\sigma M) \to Q_{(\sigma)}(M) \to Q_{(\sigma)}(M/\sigma M) \to$$
$$\to \varinjlim_{I \in \mathcal{L}(\sigma)} Ext_R^1(I, \sigma M) \to \ldots$$

From (3.12) in Chapter I, it follows that $Q_{(\sigma)}(\sigma M) = 0$. Let E the injective hull of σM, then E is σ-torsion. The exact sequence

$$0 \to \sigma M \to E \to E/\sigma M \to 0$$

induces an exact sequence

$$0 \to Q_{(\sigma)}(\sigma M) = 0 \to Q_{(\sigma)}(E) \to Q_{(\sigma)}(E/\sigma M) \to$$
$$\to \varinjlim_{I \in \mathcal{L}(\sigma)} Ext_R^1(I, \sigma M) \to \varinjlim_{I \in \mathcal{L}(\sigma)} Ext_R^1(I, E)$$

where, since $E/\sigma M$ is σ-torsion, it clearly follows that $Q_{(\sigma)}(E/\sigma M) = 0$ and where $\varinjlim_{I \in \mathcal{L}(\sigma)} Ext_R^1(I, E) = 0$, since E is injective. So,

$$\varinjlim_{I \in \mathcal{L}(\sigma)} Ext_R^1(I, \sigma M) = 0$$

and

$$Q_{(\sigma)}(M) = \varinjlim_{I \in \mathcal{L}(\sigma)} Hom_R(I, M) \cong \varinjlim_{I \in \mathcal{L}(\sigma)} Hom_R(I, M/\sigma M) \cong Q_\sigma(M).$$

This proves the assertion. $\square$

In the noetherian commutative case, this yields, in particular, for any ideal I of R and any R-module M that

$$Q_{\sigma_I}(M) = \varinjlim Hom_R(I^n, M).$$

This result is sometimes referred to as *Deligne's formula*, cf. [39].

(1.8) Proposition. *Let R be a left noetherian ring and let $\{\sigma_\alpha;\ \alpha \in A\}$ be a direct (resp. an inverse) family of stable radicals in $R-$**mod**. Then:*

(1.8.1) *if M is a left R-module, then*

$$Q_{\vee \sigma_\alpha}(M) \cong \varinjlim_{\alpha \in A} Q_{\sigma_\alpha}(M);$$

(1.8.2) *if M is a finitely generated left R-module, then*

$$Q_{\wedge \sigma_\alpha}(M) \cong \varprojlim_{\alpha \in A} Q_{\sigma_\alpha}(M).$$

Proof. If σ is an arbitrary radical in $R-$**mod** and if $f : I \to M$ is a left R-linear map, with $I \in \mathcal{L}(\sigma)$, then we denote by $[f]$ or by $[f : I \to M]$ the element in $Q_\sigma(M)$ represented by f.

To prove the first statement, first note that

$$\mathcal{L}(\bigvee_{\alpha \in A} \sigma_\alpha) = \bigcup_{\alpha \in A} \mathcal{L}(\sigma_\alpha).$$

Indeed, let $\mathcal{L} = \bigcup_{\alpha \in A} \mathcal{L}(\sigma_\alpha)$ and define for any left R-module M

$$\tau M = \{m \in M;\ \exists I \in \mathcal{L}, Im = 0\}.$$

Since the family is assumed to be direct, it is easy to see that this defines a radical τ in $R-$**mod** and that it actually coincides with $\vee_\alpha \sigma_\alpha$.

Let us now consider an arbitrary left R-module M. We have a canonical homomorphism

$$\varphi : \varinjlim_{\alpha \in A} Q_{\sigma_\alpha}(M) \to Q_\tau(M).$$

The map φ is surjective. Indeed, if $m \in Q_\tau(M)$ is represented by $f : I \to M$, for some $I \in \mathcal{L}(\vee_{\alpha \in A} \sigma_\alpha)$, then, since $\mathcal{L}(\tau) = \bigcup_{\alpha \in A} \mathcal{L}(\sigma_\alpha)$, the homomorphism also represents some $\mu \in \varinjlim Q_{\sigma_\alpha}(M)$, and obviously $\varphi(\mu) = m$.

On the other hand, φ is also injective. Indeed, if $\mu \in \varinjlim Q_{\sigma_\alpha}(M)$ is represented by some $f : I \to M$, with $I \in \mathcal{L}(\sigma_\alpha)$, and if $\varphi(\mu) = 0$, then, for some $J \subseteq I$ in $\mathcal{L}(\tau)$, we find that $f \mid_J : J \to M$ is the zero morphism. However, since $\mathcal{L}(\tau) = \bigcup_{\alpha \in A} \mathcal{L}(\sigma_\alpha)$, we may find some $\beta \in A$ such that $\sigma_\beta \geq \sigma_\alpha$ and $J \in \mathcal{L}(\sigma_\beta)$. It follows that $f \mid_J$, and hence f represents $0 \in \varinjlim Q_{\sigma_\alpha}(M)$, i.e., $\mu = 0$ and φ is injective, as well.

To prove the second statement, let us now assume M to be a finitely generated left R-module and let $\tau = \bigwedge_{\alpha \in A} \sigma_\alpha$. For every σ_α we have $\tau \leq \sigma_\alpha$, whence a homomorphism

$$\varphi : Q_\tau(M) \to \varprojlim_{\alpha \in A} Q_{\sigma_\alpha}(M).$$

In other words, if $m \in Q_{\wedge \sigma_\alpha}(M)$ is represented by some $f : I \to M$ with $I \in \mathcal{L}(\tau)$, then

$$\varphi(m) = ([f]_\alpha)_{\alpha \in A} \in \varprojlim_{\alpha \in A} Q_{\sigma_\alpha}(M) \subseteq \prod_{\alpha \in A} Q_{\sigma_\alpha}(M).$$

We claim that φ is injective. Indeed, if m is as above, and if $\varphi(m) = 0$, then there exists some $\beta \in A$, such that $[f]_\alpha = 0$ for all $\sigma_\alpha \leq \sigma_\beta$. So, for any $\sigma_\alpha \leq \sigma_\beta$. we may find some $J_\alpha \subseteq I$ in $\mathcal{L}(\sigma_\alpha)$, with $f \mid_{J_\alpha} = 0$. Let $J = \sum J_\alpha$, then $J \in \bigcap_{\sigma_\alpha \leq \sigma_\beta} \mathcal{L}(\sigma_\alpha) = \mathcal{L}(\tau)$ and $f \mid_J = 0$. This shows that $m = 0$, and hence, that φ is injective.

To prove that φ is surjective, consider

$$\mu = (m_\alpha)_{\alpha \in A} \in \varprojlim_{\alpha \in A} Q_{\sigma_\alpha}(M) \subseteq \prod_{\alpha \in A} Q_{\sigma_\alpha}(M),$$

where each m_α is represented by some $f_\alpha : I_\alpha \to M$, with $I_\alpha \in \mathcal{L}(\sigma)$. We claim that there exists some $\beta \in A$ such that $\tau M = \sigma_\gamma M$ for all $\sigma_\gamma \leq \sigma_\beta$, or equivalently, such that $M/\tau M \in \mathcal{F}_{\sigma_\beta}$. Decompose $E(M/\tau M)$ into a finite direct sum $E_1 \oplus \cdots \oplus E_n$ of indecomposable injectives, then $E_i \in \mathcal{F}_\tau$, for all $1 \leq i \leq n$, i.e., $E_i \notin \mathcal{T}_\tau = \bigcap_{\alpha \in A} \mathcal{T}_{\sigma_\alpha}$. So, there exists some $\beta \in A$ such that $E_i \notin \mathcal{T}_{\sigma_\beta}$, for all i. Since σ_β is stable, by assumption, this shows that $E(M/\tau M)$ is σ_β-torsionfree, indeed.

In the projective limit, we may, of course restrict to $\sigma_\alpha \leq \sigma_\beta$. Hence, we may as well assume M to be σ_α-torsionfree for all $\alpha \in A$. Now, since $(m_\alpha)_\alpha$ belongs to the projective limit, for all $\sigma_\delta \geq \sigma_\alpha$ in the family we have

$$[f_{\sigma_\alpha} : I_{\sigma_\alpha} \to M]_{\sigma_\delta} = [f_{\sigma_\delta} : I_{\sigma_\delta} \to M]_{\sigma_\delta},$$

i.e., there exists some $J \subseteq I_\alpha \cap I_\delta$, belonging to $\mathcal{L}(\sigma_\delta)$ and such that $f_\alpha \mid_J = f_\delta \mid_J$. Let

$$g = f_\alpha \mid_{I_\alpha \cap I_\delta} - f_\delta \mid_{I_\alpha \cap I_\delta} : I_\alpha \cap I_\delta \to M,$$

then $g \mid_J = 0$, hence g factorizes through some $h : (I_\alpha \cap I_\delta)/J \to M$. However, by the previous discussion, M is σ_α-torsionfree, whereas $(I_\alpha \cap I_\delta)/J \subseteq R/J$ is σ_α-torsion, as $J \in \mathcal{L}(\sigma_\alpha)$. So, $h = 0$, implying that $g = 0$, as well. This proves that $f_\alpha \mid_{I_\alpha \cap I_\delta} = f_\delta \mid_{I_\alpha \cap I_\delta}$. To conclude, let $I = \sum I_\alpha$, then $I \in \bigcap_{\alpha \in A} \mathcal{L}(\sigma_\alpha) = \mathcal{L}(\tau)$. Moreover, by the foregoing, we may define some homomorphism $f : I \to M$, by putting $f \mid_{I_\alpha} = f_\alpha$, for all $\alpha \in A$. If we let m denote the class of f in $Q_\tau(M)$, then clearly $\varphi(m) = \mu$. This proves that φ is surjective as well, which finishes the proof. $\qquad\square$

With a little more work, the first part of the previous result may be strengthened as follows:

(1.9) **Proposition.** *Let R be a left noetherian ring and let $\{\sigma_\alpha; \ \alpha \in A\}$ be a direct family of symmetric radicals in $R-\mathbf{mod}$. Then, for any left R-module M, there is a canonical isomorphism*

$$Q_{\vee \sigma_\alpha}(M) \cong \varinjlim_{\alpha \in A} Q_{\sigma_\alpha}(M).$$

Proof. Let $\sigma = \bigvee_{\alpha \in A} \sigma_\alpha$. If we first assume M to be σ-torsionfree, then the proof runs along the lines of that of the previous result.

In the general case, let us first assume M to be a finitely generated left R-module. Let $I = Ann_R^\ell(M)$, then $I \in \mathcal{L}^2(\sigma)$, since σ is symmetric. Again $\mathcal{L}(\sigma) = \bigcup_{\alpha \in A} \mathcal{L}(\sigma_\alpha)$, so $I \in \mathcal{L}(\sigma_\alpha)$, for some $\alpha \in A$. Let

$B = \{\beta \in A; \ \sigma_\beta \geq \sigma_\alpha\}$. If $\beta \in B$, then $\sigma_\alpha M \subseteq \sigma_\beta M \subseteq \sigma M$. As $Ann_R^\ell(M) \in \mathcal{L}^2(\sigma_\alpha)$, it follows that $\sigma M \subseteq \sigma_\alpha M$, hence that $\sigma M = \sigma_\beta M$, for any $\beta \in B$.

We thus obtain:

$$\varinjlim_{\alpha \in A} Q_{\sigma_\alpha}(M) \cong \varinjlim_{\beta \in B} Q_{\sigma_\beta}(M) \cong \varinjlim_{\beta \in B} Q_{\sigma_\beta}(M/\sigma_\beta M) \cong$$

$$\cong \varinjlim_{\beta \in B} Q_{\sigma_\beta}(M/\sigma M) \cong Q_\sigma(M/\sigma M) \cong Q_\sigma(M),$$

by the previous case and the fact that $\bigvee_{\beta \in B} \sigma_\beta = \sigma$.

Finally, if M is an arbitrary left R-module, then $M = \bigcup_\lambda M_\lambda$, for some directed family of finitely generated left R-submodules M_λ of M. Since R is assumed to be left noetherian, localization functors commute with direct unions. This allows to reduce the general case to the previous one in a rather straightforward way. $\square$

2.   The Artin-Rees property

(2.1)   In the previous Section, we have briefly examined the notion of a stable radical in $R-\mathbf{mod}$. As we will see in the present Section, it appears that this notion is tightly connected to the so-called Artin-Rees property.

For a single twosided ideal I of R, one says that I has the *Artin-Rees property* with respect to a left R-module M if for any submodule N of M and any positive integer n, there exists a positive integer r such that $I^r M \cap N \subseteq I^n N$. Clearly, the celebrated Artin-Rees Lemma, cf. [53, 99], just states that every ideal I in a noetherian commutative ring R has the Artin-Rees property with respect to *any* finitely generated R-module.

Before coming to our main results, let us mention:

(2.2) Proposition. *Let R be a ring and I a twosided ideal of R. If M is a left R-module, then the following statements are equivalent:*

(2.2.1) *I has the Artin-Rees property with respect to M;*

(2.2.2) *if N is a submodule of M, then there exists a positive integer r such that $I^r M \cap N \subseteq IN$.*

Proof. The second assertion obviously implies the first one. Conversely, let us assume (1) and prove that for any submodule N of M and any positive integer n, there exists a positive integer r such that $I^r M \cap N \subseteq I^n N$. We proceed by induction on n. If $n = 1$, the statement is obvious. Let us assume that for any submodule L of M there exists a positive integer s such that $I^s M \cap L \subseteq I^{n-1} L$. In particular, if N is a submodule of M, then so is IN and there exists a positive integer s with the property that $I^s M \cap IN \subseteq I^{n-1}(IN) = I^n N$. On the other hand, there also exists a positive integer r such that $I^r M \cap N \subseteq IN$. Thus, if $t = max\{r, s\}$, we have $I^t M \cap N \subseteq I^s M \cap IN \subseteq I^n N$. $\square$

A left R-module M is said to be a left *Artin-Rees module* if it satisfies
(2.2.2) for *any* twosided ideal I of R.

We then have the following result, due to Lesieur and Croisot:

(2.3) Proposition. [48] *Let M be a finitely generated left R-module. Then the following assertions are equivalent:*

(2.3.1) *M is a left Artin-Rees module;*

(2.3.2) *every tertiary submodule of M is primary.*

Proof. To prove that (1) implies (2), assume that M is a left Artin-Rees module and consider a tertiary submodule N of M with $P = ass_R(M/N)$, i.e., such that $P = Ann_R^\ell(M'/N)$ for some $M' \subseteq M$ containing N. Let us suppose M' to be maximal as such, then

$$M' = \{m \in M; \; Pm \subseteq N\}.$$

The Artin-Rees property then yields the existence of some positive integer n such that $P^n M \cap M' \subseteq PM' \subseteq N$. Hence $P^n M \subseteq N$, as M'/N is essential in M/N (by the maximality assumption on M'), so N is primary in M.

Conversely, to prove that (2) implies (1), assume that every tertiary submodule of M is primary and let N be a left R-submodule of M. If I is a twosided ideal of R and n is a positive integer, then P^n annihilates $N/I^n N$. So, $I^n \subseteq P$ for any $P \in Ass_R(N/I^n N)$, hence $I \subseteq P$. From (5.17) in Chapter I, it follows that there exists some $N' \subseteq M$ with $N' \cap N = I^n N$ and $Ass_R(M/N') = Ass_R(N/I^n N)$.

Let $N' = N_1 \cap \ldots \cap N_t$ be a tertiary decomposition of N' in M, say with $P_i = ass_R(M/N_i)$ for $1 \leq i \leq t$. Clearly, $P_i \in Ass_R(N/IN) = Ass_R(N/I^n N)$, so $I \subseteq P_i$. Since our assumption implies each N_i to be primary in M, we may find some positive integer m such that $P_i^m M \subseteq N_i$ for every $1 \leq i \leq t$. Hence

$$I^m M \subseteq \bigcap_{i=1}^{t} P_i^m M \subseteq \bigcap_{i=1}^{t} N_i = N',$$

so

$$I^m M \cap N \subseteq N' \cap N = I^n N,$$

which proves that M is a left Artin-Rees module, indeed. $\square$

(2.4) Lemma. *Let R be a ring and I a twosided ideal, the following statements are equivalent:*

(2.4.1) *I has the Artin-Rees property with respect to all finitely generated left R-modules;*

(2.4.2) *if N is an essential submodule of a finitely generated left R-module M such that $IN = 0$, then there exists a positive integer r such that $I^r M = 0$.*

Proof. The implication (1) $\Rightarrow$ (2) is trivial. Conversely, to prove that (2) implies (1), consider a submodule N of a finitely generated left R-module M, and let $\mathcal{S}$ denote the set of all submodules L of M such that $N \cap L = IN$. Obviously, $IN \in \mathcal{S}$, so $\mathcal{S}$ is not empty. Let L be a maximal element of $\mathcal{S}$, then $N + L/L$ is an essential submodule of M/L. Since $IN \subseteq L$, we have $I(N + L/L) = 0$, so there exists a positive integer r with the property that $I^r(M/L) = 0$, hence $I^r M \subseteq L$ and $N \cap I^r M \subseteq IN$. $\square$

(2.5) Example. Let R be a left noetherian ring and let I be a twosided ideal of R generated by elements in $N(R)$, the set of all $r \in R$ with the property that $Rr = rR$, cf. (2.1) in Chapter II, then I has the left Artin-Rees property.

Indeed, consider an essential extension $N \subseteq M$ of finitely generated left R-modules with the property that $IN = 0$, then we claim that $I^n M = 0$ for some positive integer n. For, let $a \in I \cap N(R)$ and denote by $f : M \to M$ the morphism of abelian groups defined by mapping $m \in M$ to $am \in M$. Since $a \in N(R)$, one easily verifies that $Ker(f^n)$ and $Im(f^n)$ are left R-submodules of M for each positive integer n.

The chain

$$Ker(f) \subseteq Ker(f^2) \subseteq \ldots \subseteq Ker(f^n) \subseteq Ker(f^{n+1}) \subseteq \ldots$$

stops for some positive integer n, as M is noetherian, being finitely generated over the left noetherian ring R. If $Ker(f^n) = Ker(f^{n+1})$, then $Ker(f^n) \cap Im(f^{n+1}) = 0$. But, since $N \subseteq Ker(f) \subseteq Ker(f^n)$ and since N is essential in M, it follows that $Im(f^n) = 0$, so $a^n M = 0$. Applying this to the set $\{a_1, \ldots, a_t\}$ of normal generators of I, the previous argument shows that there exists for each $1 \le i \le t$ some positive integer $n(i)$ with $a_i^{n(i)} M = 0$. For $n = \sum_{i=1}^{t} n(i)$, it then follows from $I^n \subseteq \sum_{i=1}^{t} Ra_i^{n(i)}$ that $I^n M = 0$, indeed.

In particular, if R is a commutative ring, then this implies that every ideal I of R has the Artin-Rees property. This is essentially the contents of Krull's Lemma [99].

(2.6) Example. More generally, recall that a sequence $a_1, \ldots, a_t$ of elements of R is said to be a *centralizing* resp. a *normalizing sequence* if for each $0 \le i \le t - 1$ the image of a_{i+1} in $R/\sum_{j=1}^{i} Ra_j$ is a central resp. normal element and if $\sum_{i=1}^{t} Ra_i \ne R$.

As in the previous example, let us assume R to be left noetherian. One may then prove, cf. [52], that any twosided ideal I of R with a normalizing set of generators $a_1, \ldots, a_t$ such that $Ia_1 = a_1 I$ and

$$Ia_i + \sum_{j=1}^{i-1} Ra_i = a_i I + \sum_{j=1}^{i-1} a_i R,$$

for any $1 < i \le t$, has the left Artin-Rees property. The proof of this fact may be given by induction on t and uses the previous example in the induction step. We leave details to the reader.

In particular, note that any polycentral ideal (i.e., any twosided R-ideal generated by a centralizing sequence) satisfies the Artin-Rees property. This applies in particular to rings of the form RG, where R is commutative and noetherian and G is a finitely generated nilpotent group or of the form $R \otimes_k U(\mathfrak{g})$, where k is a field, R a commutative noetherian k-algebra and $\mathfrak{g}$ a finite dimensional nilpotent Lie k-algebra, where every twosided ideal is polycentral. (For details and other examples. we refer to the literature).

The next result yields a link between the Artin-Rees property and the notion of stability:

(2.7) Proposition. *Let R be a left noetherian ring and I a twosided ideal. The following statements are equivalent:*

(2.7.1) *I has the Artin-Rees property with respect to all finitely generated left R-modules;*

(2.7.2) *I has the Artin-Rees property with respect to all left ideals of R;*

(2.7.3) *σ_I is stable.*

Proof. The implication (1) $\Rightarrow$ (2) is obvious.

To prove the implication (2) $\Rightarrow$ (3), suppose that (2) holds and let M be a σ_I-torsion left R-module. If $x \in E(M)$, then $N = Rx \cap M$ is an essential submodule of Rx, and it is σ_I-torsion. Let $N = Ry_1 + \cdots + Ry_k$, then there exists a positive integer n such that $I^n y_i = 0$ for any $1 \leq i \leq k$ and therefore $I^n N = 0$. Put $L = (N : x)$, then L is a left ideal of R, so there exists a positive integer h such that $I^h \cap L \subseteq I^n L$. Let $m \in I^h x \cap N$, then $m = jx$ for some $j \in I^h$ and as $m \in N$, it follows that actually $j \in I^h \cap L$, i.e., $j \in I^n L$. So $m = jx \in I^n Lx \subseteq I^n N = 0$. But then $I^h x \cap N = 0$, whence $I^n x = 0$, as N is essential in Rx. It thus follows that $x \in \sigma_I E(M)$, hence $E(M)$ is σ_I-torsion, as $x \in E(M)$ is arbitrary.

Let us finally prove that (3) implies (1). Let M be a finitely generated left R-module and let N be an essential submodule of M satisfying $IN = 0$. Then N is σ_I-torsion and since σ_I is stable, obviously M is σ_I-torsion. As M is finitely generated, there exists a positive integer r with the property that $I^r M = 0$. $\qquad\square$

Any twosided ideal I satisfying the equivalent conditions of the previous result is said to have the *left Artin-Rees property* .

Recall that a twosided ideal I of R is said to be *invertible*, if there exists a ring extension $R \subseteq S$ and a subset K of S, with $IK = KI = R$. Note that this subset K may then clearly be chosen to be an R-subbimodule of S containing R.

We then have the following example:

(2.8) Proposition. *If R be a left noetherian ring and I an invertible ideal of R, then I satisfies the left Artin-Rees property.*

Proof. As I is invertible, there exists a ring $S \supseteq R$ and an R-subbimodule K of S containing R, with $IK = KI = R$. We thus obtain an ascending chain of left R-modules

$$K \subseteq K^2 \subseteq K^3 \subseteq \ldots$$

Hence, for any left ideal L of R,

$$KL \cap R \subseteq K^2L \cap R \subseteq K^3L \cap R \subseteq \ldots$$

is an ascending chain of left ideals of R. As R is left noetherian, there exists a strictly positive integer n such that $K^{n-1}L \cap R = K^nL \cap R$. Since $K^n(L \cap I^n) \subseteq K^nL \cap R$, it follows that

$$L \cap I^n \subseteq I^n(K^nL \cap R) = I^n(K^{n-1}L \cap R) \subseteq IL,$$

which proves the assertion. $\Box$

The relationship between the Artin-Rees property and localizability is given by the next results:

(2.9) Lemma. *Let R be a left noetherian ring. If $P \in Spec(R)$ satisfies the left Artin-Rees property, then P satisfies the strong left second layer condition.*

Proof. Indeed, otherwise there is a finitely generated left R-module M with the property that $Ann_R^\ell(M) = Q$ is a prime ideal of R strictly contained in P and which contains an essential left R-submodule N with $PN = 0$. The Artin-Rees property implies that $P^n M = 0$ for some positive integer n. So, $P^n \subseteq Q$, hence $P \subseteq Q$, a contradiction.

$\square$

(2.10) Corollary. *Let R be a noetherian ring and P a prime ideal of R satisfying the left Artin-Rees property. Then the following assertions are equivalent:*

(2.10.1) *P is classically localizable;*

(2.10.2) *if Q is a prime ideal of R with $P \subseteq Q$ and $P \rightsquigarrow Q$, then $Q = P$.*

Proof. We only have to prove that (2) implies (1). In view of the previous Lemma, it remains to show that $\{P\}$ is left link closed. Assume there is a link $P \rightsquigarrow Q$, given by the ideal $PQ \subseteq I \subset P \cap Q = J$, i.e., $P = Ann_R^\ell(J/I)$ resp. $Q = Ann_R^r(J/I)$.

As P satisfies the left Artin-Rees property, we may find some positive integer n with
$$JP^n \subseteq J \cap P^n \subseteq PJ \subseteq I.$$

So, $P^n \subseteq Q$, hence $P \subseteq Q$. From assumption (2), it then follows that $Q = P$, indeed.

$\square$

In particular, if R is noetherian and if every prime ideal of R has the (left and right) Artin-Rees property, then, using (1.18) in Chapter I, for example, every prime ideal is easily seen to be classically localizable.

(2.11) Let σ be a radical in $R-\mathbf{mod}$. We will say that σ has the (left) *Artin-Rees property*, if for any left ideal K of R and any $I \in \mathcal{L}(\sigma)$, there exists $J \in \mathcal{L}(\sigma)$ such that $J \cap K \subseteq IK$. In particular, Krull's Lemma implies *every* radical over a noetherian commutative ring to have the Artin-Rees property.

From (2.7) it also immediately follows that for any twosided ideal I of a left noetherian ring R the radical σ_I is stable if and only if it satisfies the Artin-Rees property.

Somewhat more generally:

(2.12) Proposition. *Let R be a left noetherian ring and σ a symmetric radical $R-\mathbf{mod}$. The following statements are equivalent:*

(2.12.1) *σ is stable;*

(2.12.2) *σ has the Artin-Rees property.*

Proof. Let σ be a stable radical, K a left ideal of R and $I \in \mathcal{L}(\sigma)$, then we try to find some $J \in \mathcal{L}(\sigma)$ such that $J \cap K \subseteq IK$. We may assume that I is a twosided ideal of R. The homomorphism

$$f : K \to K/IK \hookrightarrow E(K/IK)$$

extends to R, so there exists some $e \in E(K/IK)$ with the property that $f(k) = ke$ for any $k \in K$. Since K/IK is σ-torsion, so is $E(K/IK)$, because σ is stable. Hence there exists $J \in \mathcal{L}(\sigma)$ such that $Je = 0$ and $J \cap K \subseteq Ker(f) = IK$, indeed.

Conversely, let σ be a radical satisfying the Artin-Rees property. Let M be a σ-torsion left R-module and let $E(M)$ be its injective hull. Clearly, $F = \sigma E(M)$ contains M, so it suffices to prove that F is injective. Indeed, in this case $E(M) = \sigma E(M) = F$, i.e., $E(M)$ is σ-torsion, and so σ will be stable. Let $f : K \to F$ be an arbitrary left R-linear map, where K is a left ideal of R. Since R is left noetherian, K is finitely generated, so there exists $I \in \mathcal{L}(\sigma)$ with $If(K) = 0$. By the Artin-Rees property there exists $J \in \mathcal{L}(\sigma)$ such that $J \subseteq I$ and $J \cap K \subseteq IK$, and we may assume J to be twosided. Since $f(J \cap K) \subseteq f(IK) = If(K) = 0$, we have $J \cap K \subseteq Ker(f)$. So f may be factorized through $K/J \cap K$, and it induces an R/J-homomorphism

$$\overline{f} : K/(J \cap K) \to Ann^r_{E(M)}(J).$$

Since $Ann^r_{E(M)}(J)$ is injective in $R/J-\mathbf{mod}$, we may extend $\overline{f}$ to

$$\overline{g} : R/J \to Ann^r_{E(M)}(J),$$

so the composition

$$g : R \to R/J \to Ann^r_{E(M)}(J) \subseteq F$$

extends f. □

Since the proof of the implication (1) ⇒ (2) also works without assuming σ to be symmetric, it follows:

(2.13) Corollary. *Let R be a left noetherian ring and let σ be a radical in $R-\mathbf{mod}$. If σ is stable, then it satisfies the Artin-Rees property.*

Note also:

(2.14) Corollary. *Every radical over a noetherian commutative ring is stable.*

Proof. This follows immediately from the fact that, due to Krull's Lemma, every radical over such a ring has the Artin-Rees property. □

(2.15) Example. Let us point out that over a *non*-noetherian commutative ring not every radical is stable. Indeed, let R be a commutative (non-discrete!) valuation ring with value group $Z \times Z$ with the lexicographical order. Then $Spec(R) = \{0, \mathfrak{p}, \mathfrak{m}\}$ with $0 \subset \mathfrak{p} \subset \mathfrak{m}$. In addition, clearly R is non-noetherian. The simple R-module $R/\mathfrak{m}$ defines a radical σ, by letting $\mathcal{T}_\sigma = \bigcap \mathcal{T}$, where $\mathcal{T}$ runs through all torsion classes containing $R/\mathfrak{m}$. We leave it as an easy exercise to the reader to verify that $\mathcal{L}(\sigma)$ consists of all left ideals L of R with the property that R/L has finite length, cf. [80, (VIII.2.3)], for example. Let us show that σ is not stable, by proving that $E(R/\mathfrak{m})$ is not σ-torsion. So, suppose $E(R/\mathfrak{m}) \in \mathcal{T}_\sigma$ and pick $0 \neq x \in E(R/\mathfrak{m})$. Then

$L = AnnR(x) \in \mathcal{L}(\sigma)$ and as L is artinian, there exists some positive power $\mathfrak{m}^n$ of $\mathfrak{m}$ with $\mathfrak{m}^n \subseteq L$, i.e., with $\mathfrak{m}^n x = 0$. As x was chosen arbitrarily, it thus follows that $\bigcap_{n=0}^{\infty} \mathfrak{m}^n$ annihilates $E(R/\mathfrak{m})$. However, $E(R/\mathfrak{m})$ is a cogenerator of $R-mod$, so this implies that $\bigcap_{n=0}^{\infty} \mathfrak{m}^n = 0$, cf. [80], which contradicts $0 \neq \mathfrak{p} \subseteq \bigcap_{n=0}^{\infty} \mathfrak{m}^n$. This proves our assertion.

The next result characterizes not necessarily symmetric radicals satisfying the Artin-Rees property:

(2.16) Proposition. *Let R be a left noetherian ring and σ a radical in $R-\mathrm{mod}$. The following statements are equivalent:*

(2.16.1) *σ satisfies the Artin-Rees property;*

(2.16.2) *for any essential extension $N \subseteq M$ of left R-modules, M is σ-torsion, whenever $Ann_R^{\ell}(N) \in \mathcal{L}(\sigma)$;*

(2.16.3) *$\{M \in R-\mathrm{mod};\ Ass_R(M) \subseteq \mathcal{Z}(\sigma)\} \subseteq \mathcal{T}_{\sigma}$.*

Proof. To prove that (1) implies (2), consider a radical σ satisfying the Artin-Rees property and let $N \subseteq M$ be an essential extension, such that $L = Ann_R^{\ell}(N) \in \mathcal{L}(\sigma)$. For any $m \in M$, let

$$K = Ann_R^{\ell}(m + N) = \{r \in R;\ rm \in N\}.$$

By the Artin-Rees property for L, there exists some $J \in \mathcal{L}(\sigma)$ such that $J \cap K \subseteq LK$. Then $Jm \cap N = 0$, since if $j \in J$, then $jm \in Jm \cap N$ implies that $j \in J \cap K \subseteq LK$, so $jm \in LKm \subseteq LN = 0$. Because N is essential in M, we obtain $Jm = 0$, and since $J \in \mathcal{L}(\sigma)$, this yields $m \in \sigma M$. As this holds for any $m \in M$, we find that $M \in T_{\sigma}$, indeed.

Conversely, assume (2) and let us derive (1). If K is a left ideal of R and $L \in \mathcal{L}(\sigma)$, we consider the homomorphism $f = i\pi : K \to K/LK \hookrightarrow E(K/LK)$. Since $E(K/LK)$ is injective, there exists $e \in E(K/LK)$ such that $f(r) = re$, for any $r \in K$. On the other hand, since K/LK is annihilated by $L \in \mathcal{L}(\sigma)$ we have $Ann_R^{\ell}(K/LK) \in \mathcal{L}(\sigma)$, and so, by assumption, $E(K/LK)$ is σ-torsion. Therefore $Je = 0$ for some

$J \in \mathcal{L}(\sigma)$, and $J \cap K \subseteq Ker(f) \subseteq LK$.

To prove that (2) implies (3), consider a radical σ satisfying the Artin-Rees property and let M be a left R-module such that $Ass_R(M) \subseteq \mathcal{Z}(\sigma)$. To show that M is σ-torsion, it suffices, to prove that any finitely generated submodule is σ-torsion.

Let $N \subseteq M$ be a finitely generated R-submodule, then, as in [45, 4.2.12], N is easily seen to be an essential submodule of a direct sum of the form $\oplus\{N_P;\ P \in Ass_R(N)\}$, where each N_P is P-cotertiary. Indeed, if we decompose the injective hull $E(N)$ of N into a direct sum $E(N) = \oplus_{i \in I} E_i$ of indecomposable injectives, then

$$Ass_R(N) = \{ass_R(E_i);\ i \in I\}.$$

Let E_P denote the direct sum of the E_i with $ass_R(E_i) = P$, then $E(N) = \oplus\{E_P;\ P \in Ass_R(N)\}$. Letting N_P denote for any $P \in Ass_R(N)$ the inverse image of N in E_P through the canonical projection $E(N) \to E_P$ does the trick. Thus for any $P \in Ass_R(N)$ we obtain that $Ann_{N_P}^r(P)$ is an essential left R-submodule of N_P, and that $Ann_R^\ell(Ann_{N_P}^r(P)) \supseteq P$. Using (2), we obtain $N_P \in \mathcal{T}_\sigma$. So $\oplus\{N_P;\ P \in Ass_R(N)\} \in \mathcal{T}_\sigma$, and finally $M \in \mathcal{T}_\sigma$.

Conversely, assume (3), then we claim that (2) is also valid. Indeed, if $N \subseteq M$ is an essential extension such that $Ann_R^\ell(N) \in \mathcal{L}(\sigma)$, then, since every associated prime ideal to N contains $Ann_R^\ell(N)$, it follows that $Ass_R(M) = Ass_R(N) \subseteq \mathcal{Z}(\sigma)$. Hence, from (3), it follows that $M \in \mathcal{T}_\sigma$.   □

(2.17) Corollary. *Let R be a left noetherian ring and $\{\sigma_\alpha;\ \alpha \in A\}$ a family of radicals satisfying the Artin-Rees property, then $\bigwedge_{\alpha \in A} \sigma_\alpha$ also satisfies the Artin-Rees property.*

Proof. Let $N \subseteq M$ be an essential extension of left R-modules such that $Ann_R^\ell(N) \in \mathcal{L}(\bigwedge_{\alpha \in A} \sigma_\alpha)$. Then $Ann_R^\ell(N) \in \mathcal{L}(\sigma_\alpha)$, for all $\alpha \in A$. Since every σ_α satisfies the Artin-Rees property, M is σ_α-torsion for all $\alpha \in A$, so M is $(\bigwedge_{\alpha \in A} \sigma_\alpha)$-torsion, indeed.   □

Using (2.16), we may also derive the following generalization of (2.7):

(2.18) Proposition. *Let R be a left noetherian ring and σ a symmetric radical in $R-\mathbf{mod}$. Then the following statements are equivalent:*

(2.18.1) *σ satisfies the Artin-Rees property;*

(2.18.2) *for any essential extension $N \subseteq M$ of finitely generated left R-modules, we have $Ann_R^\ell(M) \in \mathcal{L}(\sigma)$, whenever $Ann_R^\ell(N) \in \mathcal{L}(\sigma)$;*

(2.18.3) *for any extension $N \subseteq M$ of finitely generated left R-modules and any $L \in \mathcal{L}(\sigma)$, there exists $J \in \mathcal{L}(\sigma)$ such that $JM \cap N \subseteq LN$.*

Proof. To prove that (1) implies (2), note that our assumption implies M to be σ-torsion. Since M is finitely generated and σ is symmetric, it follows that $Ann_R^\ell(M) \in \mathcal{L}(\sigma)$.

Next, assume (2) and let us prove (3). Consider the family

$$\mathcal{S} = \{T \subseteq M; \ T \cap N = LN\},$$

then $LN \in \mathcal{S}$, so $\mathcal{S}$ is non-empty. As $\mathcal{S}$ is also inductive, by Zorn's lemma, there exists some maximal element $K \in \mathcal{S}$. If $(N + K)/K \cap T/K = 0$ for some $K \subseteq T \subseteq M$, then $T \cap N = K \cap N = LN$ and by maximality $T = K$. So, $(N + K)/K$ is essential in M/K. On the other hand,

$$(N + K)/K \cong N/(N \cap K) = N/LN$$

is annihilated by $L \in \mathcal{L}(\sigma)$, hence $Ann_R^\ell((N + K)/K) \in \mathcal{L}(\sigma)$. It thus follows that $J = Ann_R^\ell(M/K) \in \mathcal{L}(\sigma)$, and $JM \cap N \subseteq K \cap N = LN$. Finally, to show that (3) implies (1), note that R itself is a finitely generated left R-module, so, for any left ideal K of R and any $L \in \mathcal{L}(\sigma)$, there exists some $J \in \mathcal{L}(\sigma)$ such that $JR \cap K \subseteq LK$. Since clearly $JR \in \mathcal{L}(\sigma)$ (as $J \subseteq JR$), this proves the assertion. $\square$

We will see in (3.6) below, that the assumption that σ be symmetric may be eliminated, if we assume $\mathcal{Z}(\sigma)$ satisfies the strong left second layer condition.

Let us conclude this Section with some results showing that the Artin-Rees property is actually not very far from the notion of stability, even if we work with not necessarily symmetric radicals. These results will be used below, in the context of sheaf constructions.

(2.19) Lemma. *Let R be a noetherian ring and σ a radical in $R-\mathbf{mod}$ satisfying the Artin-Rees property. Then for any right finitely generated R-bimodule M we have $E(\sigma M) \in \mathcal{T}_\sigma$.*

Proof. From the hypotheses, it follows that σM is a right finitely generated R-subbimodule of M, so $Ann_R^\ell(\sigma M) \in \mathcal{L}(\sigma)$. Since σM is an essential submodule of $E(\sigma M)$, by (2.16) we obtain that $E(\sigma M) \in \mathcal{T}_\sigma$. $\qquad\square$

(2.20) Proposition. *Let R be a noetherian ring and σ a radical in $R-\mathbf{mod}$ satisfying the Artin-Rees property, then for any normalizing R-bimodule M, we have $E(\sigma M) = \sigma E(M)$.*

Proof. We claim that $E(\sigma M) \in \mathcal{T}_\sigma$. Since M is a normalizing R-bimodule, M can be written as $M = \sum_{i \in I} Rm_i = \sum_{i \in I} m_i R$, for some ordered index set I and some $m_i \in M$. It follows that M is the direct union of the family $M_j = \sum_{i \leq j} Rm_i$, where $j \in I$. From the noetherian assumption, it then follows that

$$E(\sigma M) = E(\sigma(\bigcup_j M_j)) = E(\bigcup_j \sigma M_j) = \bigcup_j E(\sigma M_j) \in \mathcal{T}_\sigma,$$

so, $E(\sigma M) \subseteq \sigma E(M)$. Now, assume this inclusion to be strict, for a moment, then $\sigma E(M) = E(\sigma M) \oplus F$, for some non-trivial $F \in R-\mathbf{mod}$. Pick $0 \neq f \in F \subseteq \sigma E(M)$, then there exists $r \in R$ such that $0 \neq rf \in M \cap \sigma E(M) = \sigma M \subseteq E(\sigma M)$. It thus would follow that $F \cap E(\sigma M) \neq 0$, which is a contradiction. So $E(\sigma M) = \sigma E(M)$, indeed. $\qquad\square$

3. The weak Artin-Rees property

As we will see below, due to some results of Beachy's [8], if R is an FBN ring, then the Artin-Rees property is equivalent to a much weaker property and, at the same time, closely related to the notion of biradical. In this Section, we will see how Beachy's original results generalize to rings satisfying the strong left second layer condition.

Let us start with the following remark:

(3.1) Lemma. Let σ be a radical in $R-\mathbf{mod}$, then the following statements are equivalent:

(3.1.1) for any twosided ideal K of R and any $I \in \mathcal{L}(\sigma)$, there exists some $J \in \mathcal{L}(\sigma)$ such that $KJ \subseteq IK$;

(3.1.2) for any twosided ideal K of R and any $I \in \mathcal{L}^2(\sigma)$, there exists some $J \in \mathcal{L}^2(\sigma)$ such that $KJ \subseteq IK$.

Proof. To show that (1) implies (2), consider a twosided ideal K of R and some $I \in \mathcal{L}^2(\sigma)$. Then $I \in \mathcal{L}(\sigma)$, and applying (1), we may pick some $J \in \mathcal{L}(\sigma)$ such that $KJ \subseteq IK$. Clearly, $J' = JR$ is a twosided ideal of R and since it contains J, it belongs to $\mathcal{L}^2(\sigma)$. Hence $KJ' = KJR \subseteq IKR = IK$, indeed.

Conversely, to show that (2) implies (1), let $I \in \mathcal{L}(\sigma)$. then $IR \in \mathcal{L}^2(\sigma)$. So, by (2), there exists some $J \in \mathcal{L}^2(\sigma)$ such that $KJ \subseteq IRK = IK$.

$\square$

A radical σ in $R-\mathbf{mod}$ is said to satisfy the *weak Artin-Rees property*, if it satisfies one of the equivalent conditions in the above Proposition.

Let us now assume R to be noetherian. If σ is a radical in $R-\mathbf{mod}$, then the filter $\mathcal{L}^2(\sigma)$ is multiplicatively closed. It then follows that there exists a unique (symmetric) radical σ^r in $\mathbf{mod}-R$ with the property that $\mathcal{L}^2(\sigma) = \mathcal{L}^2(\sigma^r)$.

As a first result, we may then mention:

(3.2) Proposition. *Let σ be a radical in $R-$**mod** and assume R to be noetherian. Then the following statements are equivalent:*

(3.2.1) *σ satisfies the weak Artin-Rees property;*

(3.2.2) *σ^r is ideal invariant (cf. (5.5) in the previous Chapter).*

Proof. Let us first show that (1) implies (2). Let K be a twosided ideal of R and $I \in \mathcal{L}(\sigma^r)$. Consider $I' \subseteq I$ such that $I' \in \mathcal{L}^2(\sigma^r) = \mathcal{L}^2(\sigma)$. Since σ satisfies the weak Artin-Rees property, there exists $J \in \mathcal{L}^2(\sigma) = \mathcal{L}^2(\sigma^r)$ such that $KJ \subseteq I'K \subseteq IK$. So $(K/IK)J = 0$, hence K/IK is σ^r-torsion, indeed.

Conversely, assume (2) and let us prove that σ satisfies the weak Artin-Rees property. Let K be a twosided ideal of R and $I \in \mathcal{L}^2(\sigma) = \mathcal{L}^2(\sigma^r)$. Consider as before $I' \subseteq I$ such that $I' \in \mathcal{L}^2(\sigma) = \mathcal{L}^2(\sigma^r)$. Since R is noetherian, K/IK is a left finitely generated R-bimodule. As it is σ^r-torsion, we may find some $J \in \mathcal{L}^2(\sigma^r) = \mathcal{L}^2(\sigma)$ such that $(K/IK)J = 0$. So, $KJ \subseteq IK$, which proves the assertion. $\square$

The notion of stability is related to the weak Artin-Rees property by the following result, which will be generalized in (3.5):

(3.3) Proposition. [8] *Let R be a left FBN ring and let σ denote a radical in $R-$**mod**. Then the following assertions are equivalent:*

(3.3.1) *σ is stable;*

(3.3.2) *σ satisfies the weak Artin-Rees property;*

(3.3.3) *for any $P \in \mathcal{K}(\sigma)$ and any $Q \in \mathcal{Z}(\sigma)$, there exists some $J \in \mathcal{L}(\sigma)$, with $PJ \subseteq QP$.*

Proof. It is clear that only the implication (3) $\Rightarrow$ (1) is non-trivial. So, assume (3) and let M be a finitely generated left R-module. Assume N to be a proper essential extension of σM within M. In order to derive (1), it clearly suffices to show that this leads to a contradiction. Since N properly contains σM, obviously $Ann_R^\ell(N) \notin \mathcal{L}(\sigma)$. Let $I =$

$Ann_R^\ell(T)$ be maximal in the set of all $Ann_R^\ell(N')$, where N' is a left R-submodule of N with the property that $Ann_R^\ell(N') \notin \mathcal{L}(\sigma)$. We want to show that I is a prime ideal, as this finishes the proof. Indeed, since R is a left FBN ring, by [24], R satisfies Gabriel's condition (H), i.e., if M is a finitely generated left R-module, then there exist $x_1, \ldots, x_n \in M$, such that

$$Ann_R^\ell(M) = \bigcap_{i=1}^n Ann_R^\ell(x_i).$$

Applying this to $M = T$, it follows that $I = \bigcap_{i=1}^n Ann_R^\ell(x_i)$ for some $x_1, \ldots, x_n \in T$. In particular, this yields a left R-linear map

$$\varphi : R \to T^n : r \mapsto (rx_1, \ldots, rx_n)$$

with kernel I, i.e., there is an embedding $R/I \hookrightarrow T^n$, for some positive integer n. Moreover, as N is essential over σM, it follows that $\sigma T = \sigma M \cap T$ is essential in T, hence that $\sigma(T^n)$ is essential in T^n. Hence, it follows that $\sigma(R/I) \neq 0$ and, as $I \in Spec(R)$, that $I \in \mathcal{L}(\sigma)$, a contradiction!

So, let us show I to be prime. If not, let P be a maximal right annihilator ideal of R/I, say $P = Ann_R^r(K/I)$, with K strictly containing I (hence $K \in \mathcal{L}(\sigma)$!). Clearly, P is prime. Moreover, since $KPT \subseteq IT = 0$, it follows that $PT \in \mathcal{T}_\sigma$ and, as $PT \neq 0$, since P strictly contains I, we may assume that $K = Ann_R^\ell(PT)$. Finally, $P \in \mathcal{K}(\sigma)$, for, otherwise, $P \in \mathcal{L}(\sigma)$, and from $KP \subseteq I$ and $KP \in \mathcal{L}(\sigma)$, it would follow that $I \in \mathcal{L}(\sigma)$.

Let $L = Ann_R^\ell(PT) \in \mathcal{L}(\sigma)$, then, since R is left noetherian, L contains some product $Q_1 \cdots Q_n$ of prime ideals Q_i containing L, hence belonging to $\mathcal{Z}(\sigma)$. By assumption, there exist ideals $J_i \in \mathcal{L}(\sigma)$, with the property that $PJ_i \subseteq Q_iP$, for all $1 \leq i \leq n$. So, with $J = J_1 \ldots J_n \in \mathcal{L}(\sigma)$, it follows that $PJ \subseteq Q_1 \ldots Q_n P \subseteq KP$.

Let $C = Ann_R^\ell(JT)$, then C is a proper ideal of R, (for otherwise $JT = 0$, so $J \subseteq I$ and hence $I \in \mathcal{L}(\sigma)$!) and $P \subseteq C$, since $PJT \subseteq KPT = 0$.

So, $C \in \mathcal{L}(\sigma)$, hence $CJ \in \mathcal{L}(\sigma)$. But, from $CJT = 0$, it clearly follows that $CJ \subseteq Ann_R^\ell(T) = I$, hence $I \in \mathcal{L}(\sigma)$. This contradiction proves the assertion. $\qquad\qquad\qquad\qquad\qquad\qquad\qquad\qquad\qquad\qquad$ $\square$

As a special case, let us point out that the previous result immediately shows every radical over a noetherian commutative ring to be stable. Note also that it shows that a twosided ideal I of a left FBN ring has the Artin-Rees property, if and only if for any $P \in X_R(I)$, i.e., $I \not\subseteq P$, there exists some positive integer n with $PI^n \subseteq IP$.

(3.4) Example. Generalizing an example given in [8], consider a Dedekind domain D with field of fractions K. Then the ring

$$R = \begin{pmatrix} D & 0 \\ K & K \end{pmatrix}$$

is certainly FBN and its prime spectrum consists of the ideals

$$M = \begin{pmatrix} D & 0 \\ K & 0 \end{pmatrix} \text{ resp. } P_{\mathfrak{p}} = \begin{pmatrix} \mathfrak{p} & 0 \\ K & K \end{pmatrix},$$

where $\mathfrak{p} \in Spec(D)$. Let $P = P_0$, then $P \in \mathcal{K}(\sigma_{R\backslash P})$, while $M \in \mathcal{Z}(\sigma_{R\backslash P})$. Since $MP = 0$, we clearly cannot find any $J \in \mathcal{L}(\sigma_{R\backslash P})$, with the property that $PJ \subseteq MP$. So, $\sigma_{R\backslash P}$ is not stable for any $P = P_{\mathfrak{p}}$ with $0 \neq \mathfrak{p} \in Spec(D)$.

On the other hand, since $\mathcal{K}(\sigma_{R\backslash M}) = \{M\}$ and

$$\mathcal{Z}(\sigma_{R\backslash M}) = \{P_{\mathfrak{p}}; \ \mathfrak{p} \in Spec(D)\},$$

it follows from $MP_0 = 0$ (and $P_0 \in \mathcal{L}(\sigma_{R\backslash M})$) that the radical $\sigma_{R\backslash M}$ in $R-\mathbf{mod}$ actually *is* stable.

Proposition (3.3) may be generalized as follows:

(3.5) Theorem. *Let R be a left noetherian ring and σ a radical in $R-\mathbf{mod}$. Consider the following conditions:*

(3.5.1) σ satisfies the Artin-Rees property;

(3.5.2) σ satisfies the weak Artin-Rees property;

(3.5.3) $\mathcal{Z}(\sigma)$ is left link closed;

(3.5.4) $\mathcal{K}(\sigma)$ is right link closed.

Then we have (1)$\Rightarrow$(2)$\Rightarrow$(3)$\Rightarrow$(4) and the conditions are equivalent, if $\mathcal{Z}(\sigma)$ satisfies the strong left second layer condition.

Proof. The implication (1) $\Rightarrow$ (2) is obvious.

Next, assume (2) and let us verify (3). Assume $Q \in \mathcal{Z}(\sigma)$ and consider $Q \rightsquigarrow P$, then, by definition, $Q = Ann_R^\ell(Q \cap P/I)$ resp. $P = Ann_R^r(Q \cap P/I)$, for some $QP \subseteq I \subset Q \cap P$. Applying the assumption on the ideals $Q \cap P$ and $Q \in \mathcal{L}(\sigma)$, yields the existence of some $J \in \mathcal{L}(\sigma)$ with the property that

$$(Q \cap P)J \subseteq Q(Q \cap P) \subseteq QP \subseteq I.$$

So, $J \subseteq Ann_R^r(Q \cap P/I) = P$, hence $P \in \mathcal{Z}(\sigma)$, indeed. This proves that (2) implies (3).

The equivalence of (3) and (4) follows immediately from the fact that $\mathcal{K}(\sigma)$ and $\mathcal{Z}(\sigma)$ are each other's complement within $Spec(R)$, as the ring R is left noetherian.

Finally, let us assume that $\mathcal{Z}(\sigma)$ satisfies the strong left second layer condition and suppose (3) holds true. Let L be an essential σ-torsion submodule of a finitely generated left R-module M and assume $IL = 0$, for some $I \in \mathcal{L}(\sigma)$. It is clear that every associated prime of M contains I, hence the assumption implies that

$$X = \bigcup\{cl^\ell(P); P \in Ass_R(M)\} \subseteq \mathcal{Z}(\sigma).$$

But then, it follows from (6.5) in Chapter I, that there exist $P_1, \ldots, P_n \in X$ such that $P_1 \cdots P_n M = 0$. Since $P_1, \ldots, P_n$ belong to $\mathcal{Z}(\sigma)$, so does their product, hence $Ann_R^\ell(M) \in \mathcal{L}(\sigma)$. Applying (2.16) thus finishes the proof. $\square$

Note that (3.3) may be recovered from the previous Theorem, as any left FBN ring satisfies the strong left second layer condition.

Note also that the foregoing result implies that, if R is left noetherian and if R (or, somewhat more generally $\mathcal{Z}(\sigma)$) satisfies the strong left second layer condition, that the Artin-Rees property then only depends upon $\mathcal{K}(\sigma)$ (or $\mathcal{Z}(\sigma)$) and not upon σ itself. Of course, if σ is symmetric, then σ is completely determined by $\mathcal{K}(\sigma)$.

Here is the result announced in (2.18)

(3.6) Proposition. *Let R be a left noetherian ring and σ a radical in $R-\mathbf{mod}$ with the property that $\mathcal{Z}(\sigma)$ satisfies the strong left second layer condition. Then the following statements are equivalent:*

(3.6.1) *σ satisfies the Artin-Rees property;*

(3.6.2) *for any essential extension $N \subseteq M$ of finitely generated left R-modules, we have $Ann_R^\ell(M) \in \mathcal{L}(\sigma)$, whenever $Ann_R^\ell(N) \in \mathcal{L}(\sigma)$;*

(3.6.3) *for any extension $N \subseteq M$ of finitely generated left R-modules and any $L \in \mathcal{L}(\sigma)$, there exists $J \in \mathcal{L}(\sigma)$ such that $JM \cap N \subseteq LN$.*

Proof. To prove that (1) implies (2), first note that any $P \in Ass_R(M)$ obviously contains $Ann_R^\ell(N) \in \mathcal{L}(\sigma)$, so $P \in \mathcal{Z}(\sigma)$. From the previous result, it follows that $\mathcal{Z}(\sigma)$ is left link closed, hence

$$X = \bigcup \{cl^\ell(P);\ P \in Ass_R(M)\} \subseteq \mathcal{Z}(\sigma).$$

Applying (6.5) in Chapter I, as in the previous proof, we may thus find $P_1, \ldots, P_n \in X$, hence in $\mathcal{Z}(\sigma)$, with $P_1 \ldots P_n M = 0$. So, $Ann_R^\ell(M) \in \mathcal{L}(\sigma)$, indeed.

Since the other implications may be derived in exactly the same way as in the proof of (2.18), this proves the assertion. $\square$

From (3.2) and (3.5), it immediately follows:

(3.7) Corollary. *If R is a noetherian ring and σ a radical in $R-\mathbf{mod}$ such that $\mathcal{Z}(\sigma)$ satisfies the strong right second layer condition, then* the following assertions are equivalent:

(3.7.1) σ is ideal invariant;

(3.7.2) the radical σ^r is stable in $\mathbf{mod}-R$.

(3.8) Corollary. If R is a noetherian prime pi ring and σ a radical in $R-\mathbf{mod}$, then the following statements are equivalent:

(3.8.1) σ is ideal invariant;

(3.8.2) σ is stable.

Proof. It has been proved in [15], cf. (6.10) in Chapter I, that the hypotheses imply $X \subseteq Spec(R)$ to be left link closed if and only if it is right link closed. The result thus follows trivially from the previous one, combined with (2.12) and the observation that our assumptions also imply R to satisfy the strong second layer condition and any radical σ in $R-\mathbf{mod}$ to be symmetric. $\qquad \square$

(3.9) Corollary. Let R be a noetherian ring and let σ be a radical in $R-\mathbf{mod}$ such that $\mathcal{Z}(\sigma)$ satisfies the left strong second layer condition. Then the following statements are equivalent:

(3.9.1) σ satisfies the Artin-Rees property;

(3.9.2) σ^r satisfies the Artin-Rees property;

(3.9.3) σ^r is stable;

(3.9.4) τ satisfies the Artin-Rees property, for any radical τ such that $\mathcal{K}(\tau) = \mathcal{K}(\sigma)$.

(3.10) Corollary. Let R be a left noetherian ring satisfying the left strong second layer condition and suppose σ_P satisfies the Artin-Rees property for all $P \in Spec(R)$. Then every radical in $R-\mathbf{mod}$ satisfies the Artin-Rees property. In particular R is a left classical ring, cf. [80].

Finally, note that (3.5) yields the following converse of (2.10), essentially due to Warfield, cf. [37]:

(3.11) Corollary. *Let R be a left noetherian ring and I a twosided ideal of R with the property that $V_R(I)$ satisfies the strong second layer condition. The following assertions are equivalent:*

(3.11.1) *I has the left Artin-Rees property;*

(3.11.2) *if $P, Q \in Spec(R)$ are linked through $P \rightsquigarrow Q$ and $P \in V_R(I)$, then $Q \in V_R(I)$.*

Proof. That (1) implies (2) follows as in (2.10). The converse implication follows by applying (3.5) to $\sigma = \sigma_I$. $\square$

The class of all symmetric radicals satisfying the weak Artin-Rees property has some interesting closure properties:

(3.12) Proposition. *The set of symmetric radicals which satisfy the weak Artin-Rees property is closed under taking arbitrary meets.*

Proof. Let $\{\sigma_\alpha; \ \alpha \in A\}$ be a family of symmetric radicals satisfying the weak Artin-Rees condition and let $\sigma = \bigwedge_{\alpha \in A} \sigma_\alpha$. Let K be a twosided ideal of R and $I \in \mathcal{L}(\sigma)$. Then $I \in \bigcap_{\alpha \in A} \mathcal{L}(\sigma_\alpha)$, so for each $\alpha \in A$, there exists $J_\alpha \in \mathcal{L}(\sigma)$ with $K J_\alpha \subseteq IK$. Let $J = \sum_{\alpha \in A} J_\alpha$, then $J \in \bigcap_{\alpha \in A} \mathcal{L}(\sigma_\alpha) = \mathcal{L}(\sigma)$ and $KJ \subseteq IK$, so σ satisfies the weak Artin-Rees condition, indeed. $\square$

(3.13) Lemma. *If σ and τ be two symmetric radicals in $R-$**mod** satisfying the weak Artin-Rees property, then*

$$\mathcal{L}(\sigma \vee \tau) = \{L \leq R; \ \exists I \in \mathcal{L}(\sigma), \ \exists J \in \mathcal{L}(\tau), \ such \ that \ IJ \subseteq L\}.$$

Proof. As we have seen in (4.5) in Chapter I,

$$\mathcal{L}(\sigma \vee \tau) = \{L \leq R; \ \exists I_1, .., I_n \in \mathcal{L}(\sigma) \cup \mathcal{L}(\tau), I_1 \cdots I_n \subseteq L\}.$$

Since $\mathcal{L}(\sigma)$ and $\mathcal{L}(\tau)$ are closed under taking products, and since σ and τ are symmetric, it follows that $L \in \mathcal{L}(\sigma \vee \tau)$ if and only if there exist $I_1, \ldots, I_k \in \mathcal{L}^2(\sigma)$ resp. $J_1, \ldots, J_k \in \mathcal{L}^2(\tau)$, such that $I_1 J_1 \cdots I_k J_k \subseteq L$.

It thus suffices to prove that for any $I_1, I_2 \in \mathcal{L}^2(\sigma)$ and $J_1, J_2 \in \mathcal{L}^2(\tau)$, there exist $I \in \mathcal{L}(\sigma)$ and $J \in \mathcal{L}(\sigma)$ with the property that $IJ \subseteq I_1 J_1 I_2 J_2$. Now, this is an easy consequence of the weak Artin-Rees property on τ. Indeed, for any twosided ideal I_2 of R and any $J_1 \in \mathcal{L}(\tau)$, there exists $J_1' \in \mathcal{L}(\tau)$, such that $I_2 J_1' \subseteq J_1 I_2$. So, $I_1 I_2 J_1' J_2 \subseteq I_1 J_1 I_2 J_2$, and we can take $I = I_1 I_2$, $J = J_1' J_2$. $\qquad\square$

Note It easily follows from the previous proof, that it actually suffices to assume σ *or* τ to satisfy the weak Artin-Rees property.

(3.14) Proposition. *The set of symmetric radicals which satisfy the weak Artin-Rees property is closed under taking finite joins.*

Proof. Let K be a twosided ideal of R and let $L \in \mathcal{L}(\sigma \vee \tau)$. There exist $I \in \mathcal{L}(\sigma)$ and $J \in \mathcal{L}(\tau)$, such that $IJ \subseteq L$. Applying the weak Artin-Rees property of σ, yields the existence of some $I' \in \mathcal{L}(\sigma)$ such that $KI' \subseteq IK$. In a similar way, there exists some $J' \in \mathcal{L}(\tau)$ such that $KJ' \subseteq JK$. If we let $L' = I'J'$, then we have $L' \in \mathcal{L}(\sigma \vee \tau)$, and

$$KL' = KI'J' \subseteq IKJ' \subseteq IJK \subseteq LK.$$

This proves the assertion. $\qquad\square$

4. Biradicals

(4.1) *In this Section, will assume throughout R to be a noetherian ring.*

A radical σ in $R-$**mod** is defined to be a *biradical* if there exists a radical ρ in **mod**$-R$, which satisfies $\sigma(J/I) = \rho(J/I)$ for all twosided ideals $I \subseteq J$ of R. We will then also say that the pair (σ, ρ) is a biradical.

Biradicals may be characterized in several ways:

(4.2) Lemma. *Let σ and ρ be radicals in $R-$**mod** resp. in **mod**$-R$. Then the following statements are equivalent:*

(4.2.1) *(σ, ρ) is a biradical;*

(4.2.2) *for any pair of twosided ideals $K \subseteq T$ of R, the left R-module $_R(T/K)$ is σ-torsion if and only if the right R-module $(T/K)_R$ is ρ-torsion;*

(4.2.3) *for any twosided ideal K of R, we have $\sigma(R/K) = \rho(R/K)$.*

Proof. It is easy to see that (1) implies (2) and (3). To prove that (2) implies (1), let us consider twosided ideals $K \subseteq L$ of R and assume $\sigma(L/K) = T/K$. Then T is a twosided ideal of R. Indeed, if $t \in T$ and $r \in R$, then there exists $I \in \mathcal{L}(\sigma)$ such that $It \subseteq K$, and since $(I : r) \in \mathcal{L}(\sigma)$, it follows from $(I : r)rt \subseteq K$ that $rt \in T$. Hence, since T/K is σ-torsion as a left R-module, it is also ρ-torsion as a right R-module, and thus $\sigma(L/K) = T/K \subseteq \rho(L/K)$. The other inclusion may be derived in a similar way. Finally, the equalities

$$\rho(L/K) = (L/K) \cap \rho(R/K) = (L/K) \cap \sigma(R/K) = \sigma(L/K)$$

show that (3) implies (1). $\square$

(4.3) Example. Recall from [85] that a ring is said to be *Zariski-central* if $rad(I) = rad(R(I \cap Z(R)))$, for any twosided ideal I of R. It is immediate from the definition that if the ring R is Zariski-central, then σ_I is a biradical, for every twosided ideal I.

Indeed, first note that $\sigma_I = \sigma_{R(I \cap Z(R))}$, since σ_I depends only upon the radical of I. Now, if $r \in R$ has the property that $\bar{r} \in R/K$ belongs to $\sigma_I(R/K)$ for some twosided K of R, then

$$(I \cap Z(R))^n Rr = R(I \cap Z(R))^n r \subseteq K$$

for some positive integer n. So, $\bar{r} \in \rho_I(R/K)$, where ρ_I is the right analogue of σ_I in $\mathbf{mod}-R$. By symmetry, we thus obtain $\sigma_I(R/K) = \rho_I(R/K)$, i.e., (σ_I, ρ_I) is a biradical.

(4.4) Example. [45] Let C be a (left and right) Ore set in the ring R. As before, we define radicals σ_C resp. ρ_C in $R-\mathbf{mod}$ resp. $\mathbf{mod}-R$, say by their Gabriel filters, consisting of all left (resp. right) ideals of R intersecting C non–trivially. The second Ore condition yields that $Ann_R^\ell(x) \cap C \neq \varnothing$ if and only if $Ann_R^r(x) \cap C \neq \varnothing$. Moreover, $C' = \{c + I;\ c \in C\}$ is also an Ore set in R/I for any ideal I of R. These facts immediately yield that (σ_C, ρ_C) is a biradical.

Biradicals behave nicely with respect to taking meets and joins:

(4.5) Proposition. *The set of all biradicals in $R-\mathbf{mod}$ is a complete distributive sublattice of the lattice $R-\mathbf{rad}$.*

Proof. (Note that we do not use the noetherian assumption in this proof). It suffices to verify that the meet and join of any family of biradicals is a biradical as well. So, let $\{\sigma_\alpha;\ \alpha \in A\}$ be a family of biradicals in $R-\mathbf{mod}$, where, for each $\alpha \in A$ the associated radical in $\mathbf{mod}-R$ is denoted by ρ_α.

Let $I \subseteq J$ be a pair of twosided ideals of R, then for each $\alpha \in A$, we have $\sigma_\alpha(J/I) = \rho_\alpha(J/I)$. It then follows that

$$\left(\bigwedge_{\alpha \in A} \sigma_\alpha\right)(J/I) = \bigcap_{\alpha \in A} \sigma_\alpha(J/I) = \bigcap_{\alpha \in A} \rho_\alpha(J/I) = \left(\bigwedge_{\alpha \in A} \rho_\alpha\right)(J/I),$$

hence $\bigwedge_{\alpha \in A} \sigma_\alpha$ is a biradical in $R-\mathbf{mod}$ with associated radical $\bigwedge_{\alpha \in A} \rho_\alpha$ in $\mathbf{mod}-R$, indeed.

On the other hand, let $\bigvee_{\alpha \in A} \sigma_\alpha = \sigma$ and let $\sigma(J/I) = K/I$, for the twosided ideal K of R. Since

$$J/K = (J/I)/(K/I) = (J/I)/\sigma(J/I) \in \mathcal{F}_\sigma = \bigcap_{\alpha \in A} \mathcal{F}_{\sigma_\alpha},$$

it follows that $\rho_\alpha(J/K) = \sigma_\alpha(J/K) = 0$, for all $\alpha \in A$, i.e., J/K belongs to $\bigcap_{\alpha \in A} \mathcal{F}_{\rho_\alpha} = \mathcal{F}_\rho$, where $\rho = \bigvee_{\alpha \in A} \rho_\alpha$. But then, $K/I \supseteq \rho(J/I)$, hence $\rho(J/I) \subseteq \sigma(J/I)$. The other inclusion follows by symmetry, whence equality. This proves the assertion. $\qquad \square$

The link between biradicals and the weak Artin-Rees property is given by:

(4.6) Lemma. *Let σ be a radical in $R-\mathbf{mod}$. Then:*

(4.6.1) *if σ is a biradical, then it satisfies the weak Artin-Rees property;*

(4.6.2) *conversely, if σ is symmetric and if for any $I \in \mathcal{L}^2(\sigma)$ and any twosided ideal K of R, there exists some $J \in \mathcal{L}^2(\sigma)$ such that $KJ \subseteq IK$ and $JK \subseteq KI$, then σ is a biradical.*

Proof. Let us assume (σ, ρ) to be a biradical and let us prove the first assertion. Let K be a twosided ideal of R and $I \in \mathcal{L}^2(\sigma)$, then $\rho(K/IK) = \sigma(K/IK) = K/IK$. Moreover, since R is left noetherian, in particular K/IK is finitely generated on the left. So, $J = Ann_R^r(K/IK) \in \mathcal{L}^2(\rho) = \mathcal{L}^2(\sigma)$ and $KJ \subseteq IK$, indeed.

To prove the second assertion, suppose σ to be symmetric, let $I \subseteq J$ be twosided ideals of R and let $\sigma(I/J) = K/J$, for some twosided ideal $K \subseteq I$. Since R is right noetherian, we find $L \in \mathcal{L}^2(\sigma)$, such that $LK \subseteq J$. So, by hypothesis, there exists some $L' \in \mathcal{L}^2(\sigma) = \mathcal{L}^2(\rho)$, such that $KL' \subseteq LK$, hence $KL' \subseteq J$. Therefore $\sigma(I/J) = K/J \subseteq \rho(I/J)$, which proves the assertion. $\qquad \square$

Recall that if σ is a radical in $R-\mathbf{mod}$, then the (symmetric) radical σ^r in $\mathbf{mod}-R$ is defined by $\mathcal{L}^2(\sigma) = \mathcal{L}^2(\sigma^r)$.

The following result generalizes [8, Theorem 1.6]:

(4.7) Theorem. *Let R be an FBN ring, σ a radical in $R-\mathbf{mod}$, and σ^r the associated radical in $\mathbf{mod}-R$. The following statements are equivalent:*

(4.7.1) *σ is bistable, i.e., σ and σ^r are stable;*

(4.7.2) *σ and σ^r are ideal invariant;*

(4.7.3) *σ is a biradical.*

Proof. That (1) implies (2) follows easily from (3.3), since a twosided ideal of R belongs to $\mathcal{L}(\sigma)$ if and only if it belongs to $\mathcal{L}(\sigma^r)$.

To prove that (2) implies (3), let $I \subseteq J$ be twosided ideals of R, and let us show that $_R(J/I)$ is σ-torsion if and only if $(J/I)_R$ is σ^r-torsion. If $_R(J/I)$ is σ-torsion, then we can find some $L \in \mathcal{L}^2(\sigma) = \mathcal{L}^2(\sigma^r)$, such that $LJ \subseteq I$. Since σ^r is ideal invariant by hypothesis, it follows that $LJ \in \mathcal{L}(L, \sigma^r)$, and thus $I \in \mathcal{L}(J, \sigma)$, i.e., $(J/I)_R$ is σ^r-torsion. Finally, that (3) implies (1) again follows from (3.3) and (4.6).   □

Somewhat more generally, (3.2) and (4.6) immediately yield:

(4.8) Corollary. *Let σ and ρ be symmetric radicals in $R-\mathbf{mod}$ resp. $\mathbf{mod}-R$, with the property that $\mathcal{L}^2(\sigma) = \mathcal{L}^2(\rho)$. The following statements are equivalent:*

(4.8.1) *(σ, ρ) is a biradical;*

(4.8.2) *σ and ρ satisfy the weak Artin-Rees property;*

(4.8.3) *σ and ρ are ideal invariant.*

Without the assumption that σ and ρ be symmetric, but adding the strong second layer condition, a combination of (4.6) and (3.5) yields:

(4.9) Corollary. *Assume that R satisfies the strong second layer condition and let σ and ρ be radicals in $R-\mathbf{mod}$ resp. $\mathbf{mod}-R$, such that*

$\mathcal{L}^2(\sigma) = \mathcal{L}^2(\rho)$. *Then the following statements are equivalent:*

(4.9.1) (σ, ρ) *is a biradical;*

(4.9.2) σ *and* ρ *satisfy the Artin-Rees property;*

(4.9.3) $\mathcal{K}(\sigma)$ $(= \mathcal{K}(\rho))$ *is link closed.*

Proof. It easily follows that (2) and (3) are equivalent and that (1) implies (2). So, let us prove that (2) implies (1), by arguing as in (3.2). Assume (2) holds and let I be a twosided ideal in R, then $\sigma(R/I) = K/I$ for some twosided ideal $K \supseteq I$ of R. Since R is noetherian on both sides, there exists some $J \in \mathcal{L}^2(\sigma)$ such that $J(K/I) = 0$, i.e., $JK \subseteq I$. As σ satisfies the (weak) Artin-Rees property, we may find some $J' \in \mathcal{L}^2(\sigma) = \mathcal{L}^2(\rho)$, such that $KJ' \subseteq JK$. So $(K/I)J' = 0$, i.e., $\sigma(R/I) = K/I \subseteq \rho(R/I)$. Since the other inclusion may be derived in a similar way, this proves the assertion. $\square$

If we assume σ to be symmetric as well, then we obtain:

(4.10) Corollary. *Assume that* R *satisfies the strong second layer condition and let* σ *be a symmetric radical in* $R-\mathbf{mod}$. *Then the following statements are equivalent:*

(4.10.1) (σ, σ^r) *is a biradical;*

(4.10.2) σ *and* σ^r *are stable;*

(4.10.3) σ *and* σ^r *are ideal invariant;*

(4.10.4) $\mathcal{K}(\sigma)$ *is link closed.*

Note that the previous results remain true if we only assume that $\mathcal{Z}(\sigma)$ satisfies the strong second layer condition – we will give some examples below, however, showing that we cannot do without less.

(4.11) Proposition. *Let* σ *be a radical in* $R-\mathbf{mod}$, *with the property that* $\mathcal{Z}(\sigma)$ *satisfies the strong left second layer condition. Then the following assertions are equivalent:*

(4.11.1) σ *satisfies the Artin-Rees property;*

(4.11.2) $\sigma M \subseteq \sigma^r M$ for any strongly normalizing R-bimodule M;

(4.11.3) $\sigma M \subseteq \sigma^r M$ for any centralizing R-bimodule M;

(4.11.4) σ satisfies the weak Artin-Rees property.

Proof. First we prove that (1) implies (2). Since every R-bimodule is the direct union of finitely generated R-bimodules, and σ and σ^r commute with direct unions, we may assume M to be a finitely generated strongly normalizing R-bimodule. Let $N = \sigma M$, then N is a left and right finitely generated R-subbimodule, as M is noetherian on both sides. It thus follows that there exists $I \in \mathcal{L}^2(\sigma)$ such that $IN = 0$. So, by the Artin-Rees property, it follows from (3.6) that there exists $J \in \mathcal{L}^2(\sigma) = \mathcal{L}^2(\sigma^r)$ such that $JM \cap N \subseteq IN = 0$, and since $JM = MJ$, it follows that $NJ \subseteq JM \cap N = 0$. So, N is σ^r-torsion as a right R-module and $\sigma M \subseteq \sigma^r M$.

That (2) implies (3) is obvious.

Let us now prove that (3) implies (4). As in (4.2), if L is a twosided ideal of R and $I \in \mathcal{L}^2(\sigma)$, then the R-bimodule L/IL is σ-torsion, since it is annihilated by $I \in \mathcal{L}(\sigma)$. Thus, since R/IL is a centralizing R-bimodule, we get

$$L/IL = \sigma(L/IL) = \sigma(R/IL) \cap L/IL \subseteq \sigma^r(R/IL) \cap L/IL = \sigma^r(L/IL),$$

so $L/IL \in \mathcal{T}_{\sigma^r}$. We thus obtain that

$$J = Ann_R^r(L/IL) \in \mathcal{L}^2(\sigma^r) = \mathcal{L}^2(\sigma),$$

and $LJ \subseteq IL$.

Finally, (3.5) yields that (4) implies (1), because $\mathcal{Z}(\sigma)$ satisfies the strong left second layer condition. $\square$

(4.12) Let σ and ρ be radicals in $R-\mathbf{mod}$ resp. $\mathbf{mod}-R$. We call the pair (σ, ρ) a *centralizing* resp. *strongly normalizing biradical* if $\sigma M = \rho M$, for any centralizing resp. strongly normalizing R-bimodule M.

Clearly any strongly normalizing biradical is centralizing and any centralizing biradical is a biradical.

The previous result thus immediately yields:

(4.13) Corollary. *Assume that R satisfies the strong second layer condition, and let σ and ρ be radicals in $R-\mathbf{mod}$ resp. $\mathbf{mod}-R$ such that $\mathcal{L}^2(\sigma) = \mathcal{L}^2(\rho)$. Then the following assertions are equivalent:*

(4.13.1) *(σ, ρ) is a biradical;*

(4.13.2) *(σ, ρ) is a centralizing biradical;*

(4.13.3) *(σ, ρ) is a strongly normalizing biradical.*

Note also:

(4.14) Lemma. *Let σ be a radical in $R-\mathbf{mod}$, then the following assertions are equivalent:*

(4.14.1) *σ is a centralizing biradical;*

(4.14.2) *for any positive integer n, any R-subbimodule J of R^n and any $I \in \mathcal{L}^2(\sigma)$, there exists some $I' \in \mathcal{L}^2(\sigma)$, with the property that $I'J \subseteq JI$ and $JI' \subseteq IJ$.*

Proof. Let $\rho = \sigma^r$. To prove that (1) implies (2), first note that (1) obviously yields that (σ, ρ) is a biradical. On the other hand, let J and I be as in (2), then $M = R^n/JI$ is a centralizing R-bimodule. Let $M' = J/JI$, then clearly $M' \in \mathcal{T}_\rho$ and from $\sigma M = \rho M$, it follows that $\sigma M' = M'$. Now, if M' is generated on the right by $\{m_1, \ldots, m_n\}$, then

$$I' = Ann_R^\ell(M) \supseteq \bigcap_{i=1}^n Ann_R^\ell(m_i) \in \mathcal{L}(\sigma),$$

whence $I' \in \mathcal{L}(\sigma)$ and $I'J \subseteq JI$, as $I'M = 0$. The other inclusion may be derived in a similar way.

Conversely, to prove that (2) implies (1), let M be a centralizing R-bimodule, which we may obviously assume to be finitely generated. It is then clear that M is (up to isomorphism) of the form R^n/L, for some positive integer n and some R-subbimodule L of R^n. Let us verify that $\sigma M = \rho M$. If $\sigma M = J/L$ for some R-subbimodule J of R^n, then we may find some $I \in \mathcal{L}^2(\sigma)$ with $I(\sigma M) = 0$, i.e., $IJ \subseteq L$. Choose $I' \in \mathcal{L}^2(\sigma) = \mathcal{L}^2(\rho)$, such that $JI' \subseteq IJ$, then $(\sigma M)I' = 0$, whence $\sigma M \subseteq \tau M$. The other inclusion may again be proved in a similar way.

$\square$

Let us conclude this Section with some examples.

(4.15) Example. Let σ_{coart} be the radical in $R-\mathbf{mod}$, whose Gabriel filter consists of all left ideals of R, which contain a *co-artinian twosided ideal* of R, i.e., a twosided ideal I of R with the property that the quotient ring R/I is left artinian. (For example, denoting by J the Jacobson radical of R, then clearly $\sigma_{coart} = \sigma_J$, if R/J is artinian).

We claim that σ_{coart} is a centralizing (even strongly normalizing) biradical.

Indeed, it clearly suffices to find some radical ρ in $\mathbf{mod}-R$, with the property that $\sigma M = \rho M$, for *any* finitely generated R-bimodule M, where $\sigma = \sigma_{coart}$. Let us choose ρ to be the right analogue of σ, i.e., let $\mathcal{L}(\rho)$ be the Gabriel filter of all *right* R-ideals, which contain some co-artinian twosided ideal.

Let $N = \sigma M$, then there exists some co-artinian twosided ideal I of R with $IN = 0$, i.e., N is an artinian left R-module. From [45, 5.2.10.], it then follows that N is also artinian as a right R-module. If N is generated by $\{n_i;\ 1 \leq i \leq n\}$ as a left R-module, then $\bigcap_{i=1}^{n} Ann_R^r(n_i) = Ann_R^r(N)$ and

$$R/Ann_R^r(n_i) \cong n_i R \subseteq N$$

in $\mathbf{mod}-R$. In particular, $R/Ann_R^r(n_i)$ is an artinian right R-module, for each $1 \leq i \leq n$, hence so is $R/Ann_R^r(N)$. It follows that $Ann_R^r(N) \in \mathcal{L}^2(\rho)$ and $N \in \mathcal{T}_\rho$.

So, $\sigma M = N \subseteq \rho M$, whence equality, as the other inclusion may be deduced in a similar way.

(4.16) Example. [63] Let k be a field and let $\mathcal{L}$ denote the set of all left ideals of a k-algebra R which are of *finite codimension* in R. It is easy to see that $\mathcal{L}$ is a *bounded filter*, (i.e., that every left ideal in $\mathcal{L}$ contains a twosided ideal also in $\mathcal{L}$) and that the product of two left ideals in $\mathcal{L}$ also belongs to $\mathcal{L}$. This implies that $\mathcal{L}$ is the Gabriel filter associated to a symmetric radical, which will be denoted by σ_{cof}.

We claim that $\sigma = \sigma_{cof}$ is a centralizing biradical.

Indeed, let M be a centralizing R-bimodule generated by

$$\{m_i;\ 1 \leq i \leq n\} \subseteq M^R.$$

So, $\alpha x = x\alpha$ for any $\alpha \in k$ and any $m \in M = RM^R$. It then obviously follows that for any R-subbimodule N of M, we have $dim_k({}_kN) < \infty$ if and only if $dim_k(N_k) < \infty$. In particular, if ρ is the right analogue of σ in $\mathbf{mod}-R$, then we obtain that $\sigma M = \rho M$, which proves the assertion.

(4.17) Example. As pointed out before, the notions of a (centralizing) biradical and a bistable radical do not coincide. Let us show this by an example, due to J. Mulet [63].

Consider a Lie algebra $\mathfrak{g}$ over some field k and let $U(\mathfrak{g})$ be its enveloping algebra. There exists a k-anti-automorphism ε of $U(\mathfrak{g})$, i.e., a k-linear endomorphism ε of $U(\mathfrak{g})$ with the property that $\varepsilon(1) = 1$ and $\varepsilon(ab) = \varepsilon(b)\varepsilon(a)$ for any $a,\ b \in U(\mathfrak{g})$. The morphism ε is given by

$$\varepsilon(x_{i_1} \ldots x_{i_n}) = (-1)^n x_{i_n} \ldots x_{i_1},$$

where $\{x_1, \ldots, x_m\}$ is a k-basis for $\mathfrak{g}$.

It thus easily follows, in particular, that σ_{cof} is stable if and only if $\rho_{cof} = \sigma_{cof}^r$ (the right analogue of σ_{cof}) is stable.

We may now formulate:

(4.18) Proposition. [63] *Let $\mathfrak{g}$ be a Lie algebra of finite dimension over some field k of characteristic 0 and let $R = U(\mathfrak{g})$. Then the following assertions are equivalent:*

(4.18.1) *σ_{cof} is (bi)stable;*

(4.18.2) *$\mathfrak{g}$ is solvable.*

Proof. Assume first $\mathfrak{g}$ to be solvable, then it is well known that R satisfies the strong second layer condition, cf. [45, pp. 225-226]. So, σ_{cof} is stable, being a biradical.

Conversely, assume that $\mathfrak{g}$ is not solvable and let $\mathfrak{r}$ be the radical of $\mathfrak{g}$ (the maximal solvable ideal of $\mathfrak{g}$). Then $\overline{\mathfrak{g}} = \mathfrak{g}/\mathfrak{r}$ is a nonzero semisimple Lie algebra over k, with the property that $\overline{R} = U(\overline{\mathfrak{g}}) \cong R/I$, with $I = \mathfrak{r}U(\mathfrak{g})$. From (4.12), it then follows that

$$\mathcal{L}(\overline{\sigma_{cof}}) = \{\overline{J}; \ J \in \mathcal{L}(\sigma_{cof})\},$$

where $\overline{J}$ is the canonical image of J in R/I. It follows that

$$\mathcal{L}(\overline{\sigma_{cof}}) = \{\overline{J}; \ dim_k(\overline{R}/\overline{J}) < \infty\}.$$

So, we may assume $\mathfrak{g}$ to be a nonzero semisimple Lie algebra with the property that $\sigma = \sigma_{cof}$ is stable.

Let $P = \mathfrak{g}U(\mathfrak{g}) = U(\mathfrak{g})\mathfrak{g}$, the twosided ideal of R generated by $\mathfrak{g}$. Then $dim_k(R/P) = 1$, i.e., $P \in \mathcal{T}_\sigma$. Let $E = E(R/P)$ denote an injective hull of R/P, then $E \in \mathcal{T}_\sigma$, as σ is assumed to be stable. If $R/P \neq E$, choose a finitely generated R-submodule of E with the property that $R/P \subseteq M$. As $M \in \mathcal{T}_\sigma$, it follows that $dim_k(M) < \infty$. Weyl's Theorem [25] then implies M to be a semisimple left R-module. In particular, R/P is a direct factor of M, which contradicts the fact that R/P is an essential left R-submodule of M. It thus follows that $R/P = E$, i.e., that R/P is an injective left R-module.

The exact sequence of left R-modules

$$0 \to P \to R \to R/P \to 0$$

induces, by the injectivity of R/P, an exact sequence

$$0 \to Hom_R(R/P, R/P) \xrightarrow{i} Hom_R(R, R/P) \to Hom_R(P, R/P) \to 0.$$

As i is an isomorphism, it follows that $Hom_R(P, R/P) = 0$. On the other hand, P is a nonzero ideal of the noetherian prime ring R, as $0 \neq \mathfrak{g} \subseteq P$. Goldie's Theorem then yields a monomorphism $j : R \hookrightarrow P$ of left R-modules. The canonical surjection $\pi : R \to R/P$ thus extends to a morphism $\pi' : P \to R/P$ with $\pi'j = \pi$. Since $\pi' \neq 0$, this yields a contradiction. $\square$

Chapter IV

COMPATIBILITY AND SHEAVES

Why do we care about sheaves and, in particular, structure sheaves at all? Well, essentially because we would like some kind of dictionary, as complete as possible, between commutative algebra and algebraic geometry, i.e., we want to be able to apply methods stemming from commutative algebra to attack problems in algebraic geometry and, conversely, to apply our geometric intuition to solve problems within the field of commutative algebra.

In fact, associating to any commutative ring R its affine scheme $Spec(R)$ and being able to recover R as the ring of global sections of the structure sheaf of $Spec(R)$ provides us exactly with such a dictionary and allows to interpret commutative algebra as a *local* version of algebraic geometry.

In noncommutative algebra, we wish to realize something similar: we want to associate to any ring R a geometrically well-behaved object, which corresponds bijectively to R and which allows us to apply geometric methods to study R. At this moment, this program is clearly far too ambitious, (although large parts of it have been realized for special classes of rings, cf. [87], for example). What we *will* do in this Chapter, is to provide a start towards the realization of the above goals, i.e., we will construct in a functorial way reasonably nice geometric objects associated to noncommutative rings and provide some first notions towards their further study. (We will certainly come back to these topics later).

So, what do we want?

In the first place, we want to associate to R a (preferably quasicompact) topological space $Spec(R)$. As indicated before, even this point is far from trivial – there are several natural choices for $Spec(R)$, each one with its own legion of supporters. Since most of the rings we are dealing with have large numbers of twosided ideals, the prime spectrum, i.e., the set of all twosided prime ideals of R endowed with the Zariski topology (or some decent subtopology) appears to be an appropriate candidate.

Now, just as in the commutative case, $Spec(R)$ itself is clearly insufficient to represent the ring R, as highly non-isomorphic rings may well have the same associated prime spectrum. So, for this reason (and others – like the requirement of being able to glue together affine schemes to arbitrary schemes), we want to endow the topological space $Spec(R)$ with a structure sheaf, which yields back the ring R by taking global sections. This sheaf should have accessible stalks (obtained by localization, for example) and should behave functorially. Of course, as we have amply discussed in the previous Chapters, this will impose serious (unavoidable) restrictions on the ring homomorphisms we consider.

As we will see below, the previous Chapters contain just the right amount of material necessary to realize these aims. In fact, it appears that the notions of compatibility (to which the first Section of this Chapter is devoted) and stability (in one of its many disguises) will play a central role in this development.

1. Sheaves

(1.1) Let us briefly recollect here some generalities concerning sheaves on topological spaces. The reader familiar with sheaf theory may skip this Section or restrict to taking a quick glance at terminology and notation. For details or proofs, we refer to [28, 29, 30, 38, 39, 82, 89, et al].

Let X be an arbitrary topological space. Denote by $Open(X)$ the set of all open subsets of X and let B be a basis for the topology of X – in particular, we could take $B = Open(X)$.

We may make B into a category, letting its objects be the open subsets of X belonging to B and by defining its morphisms as follows:

$$\begin{cases} Hom(U,V) = \varnothing & \text{if } U \not\subseteq V \\ Hom(U,V) = \{i_{U,V}\} & \text{if } U \subseteq V, \end{cases}$$

where U and V belong to B and where $i_{U,V} : U \hookrightarrow V$ denotes the canonical inclusion, when $U \subseteq V$.

(1.2) A *presheaf* (of abelian groups) on B is a contravariant functor

$$\mathcal{F} : B \to (abelian\ groups),$$

where $(abelian\ groups)$ denotes the category of abelian groups. If B is the whole set $Open(X)$, then we will just speak of a presheaf on X.

In other words, a presheaf $\mathcal{F}$ on B associates to any open subset U in B an abelian group $\mathcal{F}(U)$ and to any inclusion $V \subseteq U$ of open sets in B a group homomorphism $\mathcal{F}_V^U : \mathcal{F}(U) \to \mathcal{F}(V)$, with the property that $\mathcal{F}_U^U$ is the identity on $\mathcal{F}(U)$ for any $U \in B$ and that $\mathcal{F}_W^V \mathcal{F}_V^U = \mathcal{F}_W^U$, for any $W \subseteq V \subseteq U$ in B.

The elements of $\mathcal{F}(U)$ are usually called *sections* of $\mathcal{F}$ over U; if $U = X$, then we speak of *global sections*. If no ambiguity arises, we will sometimes write $s \mid V$ for $\mathcal{F}_V^U(s)$, whenever $V \subseteq U$ in B and s is a section of $\mathcal{F}$ over U.

We denote by $\mathcal{P}(B)$ resp. $\mathcal{P}(X)$ the category of presheaves on B resp. on X, where morphisms are just natural transformations between the corresponding functors.

In other words, a morphism $\varphi : \mathcal{E} \to \mathcal{F}$ between presheaves consists of a family of morphisms $\varphi(U) : \mathcal{E}(U) \to \mathcal{F}(U)$, one for each $U \in B$, such that the diagram

$$\begin{array}{ccc} \mathcal{E}(U) & \xrightarrow{\mathcal{E}_V^U} & \mathcal{E}(V) \\ {\scriptstyle\varphi(U)}\downarrow & & \downarrow{\scriptstyle\varphi(V)} \\ \mathcal{F}(U) & \xrightarrow[\mathcal{F}_V^U]{} & \mathcal{F}(V) \end{array}$$

commutes for every $U \supseteq V$ in B.

(1.3) It is fairly easy to see that the category $\mathcal{P}(B)$ is a Grothendieck category for any basis B of open sets of X. In fact, the characteristic properties for Grothendieck categories may be verified directly on sections. In particular, a sequence of presheaves

$$0 \to \mathcal{F}' \to \mathcal{F} \to \mathcal{F}'' \to 0$$

is exact in $\mathcal{P}(B)$ if and only if for every $U \in B$ the induced sequence of abelian groups

$$0 \to \mathcal{F}'(U) \to \mathcal{F}(U) \to \mathcal{F}''(U) \to 0$$

is exact.

(1.4) Consider the following properties:

(1.4.1) *for every $U \in B$, every open covering $\{U_\alpha \in A\}$ of U by open subsets in B and every couple of sections s, $t \in \mathcal{F}(U)$, it follows from $s \mid U_\alpha = t \mid U_\alpha$ for all $\alpha \in A$, that $s = t$;*

(1.4.2) *for every $U \in B$, every open covering $\{U_\alpha;\ \alpha \in A\}$ of U by open subsets in B and any family $\{s_\alpha \in \mathcal{F}(U_\alpha);\ \alpha \in A\}$ with the property that $s_\alpha \mid V = s_\beta \mid V$ for every $V \subseteq U_\alpha \cap U_\beta$ belonging to B, there exists a (unique) $s \in \mathcal{F}(U)$ with $s \mid U_\alpha = s_\alpha$ for all $\alpha \in A$.*

If $\mathcal{F}$ is a presheaf on B satisfying (1.4.1), then $\mathcal{F}$ is said to be *separated* and if $\mathcal{F}$ also satisfies (1.4.2), then we call $\mathcal{F}$ a *sheaf* on B.

Clearly, if $\mathcal{F}$ is a presheaf on B, then $\mathcal{F}$ is a sheaf if and only if for any $U \in B$ and any open covering $\{U_\alpha;\ \alpha \in A\}$ of U by open subsets in B, the diagram

$$0 \longrightarrow \mathcal{F}(U) \longrightarrow \prod_{\alpha \in A} \mathcal{F}(U_\alpha) \rightrightarrows \prod_{\alpha,\beta \in A} \prod_{V \subseteq U_\alpha \cap U_\beta} \mathcal{F}(V)$$

is exact (i.e., an equalizer diagram).

Of course, if $B = Open(X)$ and if open subsets of X are quasicompact, then it is fairly easy to see that one may reduce the verification of the sheaf property for $\mathcal{F}$ to that of the exactness of sequences of the form

$$0 \longrightarrow \mathcal{F}(U \cup V) \longrightarrow \mathcal{F}(U) \times \mathcal{F}(V) \rightrightarrows \mathcal{F}(U \cap V),$$

where U and V are open in X.

(1.5)   Denote by $\mathcal{S}(B)$ resp. $\mathcal{S}(X)$ the full subcategory of $\mathcal{P}(B)$ resp. $\mathcal{P}(X)$ consisting of sheaves on B resp. X.

If $\mathcal{E}$ is a sheaf on X, then it is clear that we may also view $\mathcal{E}$ as a sheaf on B. Conversely, any sheaf on B uniquely extends to a sheaf on X, i.e., the categories $\mathcal{S}(B)$ and $\mathcal{S}(X)$ are linked through the adjoint equivalences

$$res_B : \mathcal{S}(X) \to \mathcal{S}(B)$$

resp.

$$ext^X : \mathcal{S}(B) \to \mathcal{S}(X).$$

For example, $ext^X(\mathcal{E}) = \mathcal{F}$ is defined on any $U \subseteq X$ by putting

$$\mathcal{F}(U) = \varinjlim H^0(\mathfrak{U}, \mathcal{E}),$$

where $\mathfrak{U}$ runs over all open coverings of U and where, for every open covering $\mathfrak{U} = \{U_\alpha;\ \alpha \in A\}$, the group $H^0(\mathfrak{U}, \mathcal{E})$ fits into the exact diagram

$$0 \longrightarrow H^0(\mathfrak{U}, \mathcal{E}) \longrightarrow \prod_{\alpha \in A} \mathcal{E}(U_\alpha) \rightrightarrows \prod_{\alpha,\beta \in A} \prod_{V \subseteq U_\alpha \cap U_\beta} \mathcal{E}(V).$$

(1.6) Actually, this construction also works, if we just assume $\mathcal{E}$ to be a *presheaf* on B. We then define $L_B : \mathcal{P}(B) \to \mathcal{S}(X)$ by putting for every open subset U in X and every presheaf $\mathcal{E}$ on B

$$(L_B\mathcal{E})(U) = \varinjlim H^0(\mathfrak{U}, \mathcal{E}),$$

where, as above, $\mathfrak{U}$ runs over the open coverings of U. One may show that if $\mathcal{E}$ is a presheaf on B, then $L_B\mathcal{E}$ is only separated, in general, and that $L_B\mathcal{E}$ is a sheaf, whenever we start from a separated presheaf $\mathcal{E}$.

We thus obtain a functor, the so-called *sheafification* functor

$$\mathbf{a}_B : \mathcal{P}(B) \to \mathcal{S}(B),$$

by applying $res_B L_B$ twice. It is readily verified that $\mathbf{a}_B$ is left adjoint to the inclusion

$$i_B : \mathcal{S}(B) \hookrightarrow \mathcal{P}(B).$$

If $B = Open(X)$, then we will just write $\mathbf{a}$ or $\mathbf{a}_X$ for the functor $\mathbf{a}_{Open(X)}$, the left adjoint of the inclusion $i = i_X : \mathcal{S}(X) \hookrightarrow \mathcal{P}(X)$.

(1.7) Just as the category of presheaves, the category $\mathcal{S}(B)$ of sheaves on B is a Grothendieck category for any basis B of open sets of X. Here exactness is slightly more complicated than in the presheaf case, however.

Indeed, let us define for any $\mathcal{F} \in \mathcal{P}(B)$ and any $x \in X$ by

$$\mathcal{F}_x = \varinjlim\{\mathcal{F}(U);\ U \in B,\ x \in U\}$$

the *stalk* of $\mathcal{F}$ at x. Then $\mathcal{F}_x$ is an abelian group and for any $U \in B$, there is a canonical group homomorphism $\mathcal{F}_x^U : \mathcal{F}(U) \to \mathcal{F}_x$. Note also, that if $\mathcal{F}$ is a presheaf on B, then $(\mathbf{a}_B\mathcal{F})_x = \mathcal{F}_x$, for any $x \in X$. A sequence of sheaves

$$0 \to \mathcal{F}' \to \mathcal{F} \to \mathcal{F}'' \to 0$$

will now be exact in $\mathcal{S}(B)$ if and only if for any $x \in X$

$$0 \to \mathcal{F}'_x \to \mathcal{F}_x \to \mathcal{F}''_x \to 0$$

is an exact sequence of abelian groups. Of course, a sequence of sheaves which is exact in $\mathcal{P}(B)$ is also exact in $\mathcal{S}(B)$, but not conversely. Finally, it is easy to see that the sheafification functor $\mathbf{a}_B : \mathcal{P}(B) \to \mathcal{S}(B)$ is exact. Since it is left adjoint to the inclusion $i : \mathcal{S}(B) \hookrightarrow \mathcal{P}(B)$, it is a *reflector*, in the sense of [60, 78, 80, 86, 89, et al].

(1.8) Let $f : X \to Y$ denote a continuous map between the topological spaces X and Y, then for any presheaf $\mathcal{E}$ on X, we may define a presheaf $f_*\mathcal{E}$ on Y by putting

$$(f_*\mathcal{E})(V) = \mathcal{E}(f^{-1}(V)),$$

for any open subset V of Y with the obvious restriction morphisms. This yields a contravariant functor $f_* : \mathcal{P}(X) \to \mathcal{P}(Y)$, the so-called *direct image functor*, which restricts to a functor $f_* : \mathcal{S}(X) \to \mathcal{S}(Y)$. Note also that if $g : Y \to Z$ is a second continuous map, then $(gf)_* = g_*f_*$.

On the other hand, if $\mathcal{F}$ is a presheaf on Y, then one may define a presheaf $f_p^{-1}\mathcal{F}$ on X, by putting

$$(f_p^{-1}\mathcal{F})(U) = \varinjlim_{f(U) \subseteq V} \mathcal{F}(V),$$

for any open subset U of X, the inductive limit being taken over all open subsets V of Y containing $f(U)$ and with obvious restriction morphisms.

In general, $f_p^{-1}\mathcal{F}$ is not a sheaf, even when $\mathcal{F}$ is. We let $f^{-1}\mathcal{F}$ (or $f_s^{-1}\mathcal{F}$, when ambiguity arises) denote the associated sheaf $\mathbf{a}_X f_p^{-1}\mathcal{G}$. This yields a covariant functor

$$f^{-1} : \mathcal{S}(Y) \to \mathcal{S}(X),$$

called the *inverse image functor*. Note also that if $g : Y \to Z$ is a second continuous map, then $(gf)^{-1} = f^{-1}g^{-1}$.

It is easy to verify that $f^{-1} : \mathcal{S}(Y) \to \mathcal{S}(X)$ is left adjoint to $f_* : \mathcal{S}(X) \to \mathcal{S}(Y)$.

(1.9) Example. As an example, assume there is given an embedding $i : X \hookrightarrow Y$, i.e., assume that the topology on X is induced from the topology on Y, through the map i. In this case, for any sheaf $\mathcal{F}$ on Y, we will denote the sheaf $i^{-1}\mathcal{F}$ by $\mathcal{F}|X$ and call it the sheaf on X *induced* by $\mathcal{F}$ (through i). It is easy to see that for any x in X, we then have $(\mathcal{F}|X)_x = \mathcal{F}_x$.

Note also that if $X \subseteq Y$ is open, then $(\mathcal{F}|X)(U) = \mathcal{F}(U)$ for any open subset U of X.

(1.10) A (pre)sheaf of abelian groups $\mathcal{A}$ on some basis B of the topology on X is said to be a *(pre)sheaf of (commutative) rings* on B, if $\mathcal{A}(U)$ is endowed with a (commutative) ring structure for each $U \in B$ and if the corresponding restriction maps are ring homomorphisms. We will then also say that the couple $(B, \mathcal{A})$ is a *(commutative) (pre)ringed space*. In this case, it is easy to see that the stalks of $\mathcal{A}$ are rings as well and that the stalk maps are ring homomorphisms. If $B = Open(X)$, then we say that $\mathcal{A}$ is a (pre)sheaf of (commutative) rings on X and we speak of the (commutative) (pre)ringed space $(X, \mathcal{A})$.

If $\mathcal{A}$ is a (pre)sheaf of rings on X and if $f : X \to Y$ is a continuous map, then it is easy to see that $f_*\mathcal{A}$ is a (pre)sheaf of rings on Y. This allows us to define a morphism of (pre)ringed spaces $(X, \mathcal{A}) \to (Y, \mathcal{B})$ to be a couple (f, θ), where $f : X \to Y$ is a continuous map and where $\theta : \mathcal{B} \to f_*\mathcal{A}$ is a morphism of sheaves of rings (defined in the obvious way).

(1.11) Let $(B, \mathcal{A})$ be a (pre)ringed space. A (pre)sheaf $\mathcal{M}$ on B is said to be a *(pre)sheaf of left $\mathcal{A}$-modules*, if $\mathcal{M}(U) \in \mathcal{A}(U)-\mathbf{mod}$ for each $U \in B$ and if for each $V \subseteq U$ in B the restriction maps $\mathcal{M}_V^U$ is left semilinear with respect to the ring homomorphism $\mathcal{A}_V^U$. It is then also easy to see that for each $x \in X$, the stalk $\mathcal{M}_x$ is endowed with a canonical left $\mathcal{A}_x$-module structure.

One may now define in the obvious way $\mathcal{P}(B, \mathcal{A})$ resp. $\mathcal{S}(B, \mathcal{A})$, the category of presheaves resp. sheaves of left $\mathcal{A}$-modules on B. The categories $\mathcal{P}(X, \mathcal{A})$ and $\mathcal{S}(X, \mathcal{A})$ are defined similarly.

In particular, note also that for any preringed space, the sheafification functor $\mathbf{a}_B : \mathcal{P}(B) \to \mathcal{S}(B)$ restricts to an (exact) functor $\mathbf{a}_B : \mathcal{P}(B, \mathcal{A}) \to \mathcal{S}(B, \mathbf{a}_B\mathcal{A})$, which is left adjoint to the inclusion.

(1.12) Let $(f, \theta) : (X, \mathcal{A}) \to (Y, \mathcal{B})$ be a morphism of ringed spaces and let $\mathcal{F} \in \mathcal{S}(X, \mathcal{A})$. Then, clearly, $f_*\mathcal{F} \in \mathcal{S}(Y, f_*\mathcal{A})$, hence also $f_*\mathcal{F} \in \mathcal{S}(Y, \mathcal{B})$, through θ. This defines a functor

$$f_* : \mathcal{S}(X, \mathcal{A}) \to \mathcal{S}(Y, \mathcal{B}),$$

which is easily seen to be left exact.

On the other hand, if $\mathcal{G} \in \mathcal{S}(Y, \mathcal{B})$, then $f^{-1}\mathcal{G} \in \mathcal{S}(X, f^{-1}\mathcal{B})$. Since f_* is adjoint to f^{-1}, the morphism $\theta : \mathcal{B} \to f_*\mathcal{A}$ corresponds to a morphism (of sheaves of rings) $\theta' : f^{-1}\mathcal{B} \to \mathcal{A}$. Defining the tensor product of sheaves in the obvious way (constructing it locally on sections and sheafifying afterwards), this allows us to define

$$f^*\mathcal{G} = f^{-1}\mathcal{G} \otimes_{f^{-1}\mathcal{B}} \mathcal{A} \in \mathcal{S}(X, \mathcal{A}).$$

This defines a functor

$$f^* : \mathcal{S}(Y, \mathcal{B}) \to \mathcal{S}(X, \mathcal{A}),$$

which is easily seen to be exact and left adjoint to f_*.

(1.13) Example. If $i : X \hookrightarrow Y$ is injective and if the topology on X is induced by that on Y through i, then any sheaf of rings $\mathcal{B}$ on Y yields a sheaf of rings $\mathcal{A} = \mathcal{B}|Y = i^{-1}\mathcal{B}$ on X. In this case, it is easy to see that the functor i^* defined above coincides with the previously introduced functor i^{-1}. In particular, for any $\mathcal{G} \in \mathcal{S}(Y, \mathcal{B})$, we will thus continue writing $\mathcal{G}|X$ for the induced sheaf of $\mathcal{A}$-modules on X.

2. The commutative theory

(2.1) In this Section, R will always denote a commutative ring (with unit). If S and T are multiplicatively closed subsets of R, then it is easy to see (and well-known, of course), that $S^{-1}T^{-1}M = T^{-1}S^{-1}M$, for any R-module M. Somewhat more generally, if σ and τ are two radicals, which possess Goldman's property (T), cf. (3.17) in Chapter I, then the associated localization functors Q_σ and Q_τ commute, i.e., for any R-module M, we have $Q_\sigma Q_\tau(M) = Q_\tau Q_\sigma(M)$. Indeed, in this case we have:

$$Q_\sigma(Q_\tau(M)) = Q_\sigma(R) \otimes (Q_\tau(R) \otimes M) = Q_\tau(R) \otimes (Q_\sigma(R) \otimes M)$$
$$= Q_\tau(Q_\sigma(M)).$$

In this Section, we will develop some machinery, which allows us to verify that under certain circumstances the previous commutativity property still holds, without having to assume σ or τ to have property (T). Somewhat surprisingly, it appears that the most elegant way to do this is by introducing some sheaf theoretic machinery.

(2.2) In classical algebraic geometry, one starts from a commutative ring R and the associated topological space $Spec(R)$ of all prime ideals of R, endowed with the Zariski topology, i.e., with open subsets of the form

$$X_R(S) = \{\mathfrak{p} \in Spec(R); S \not\subseteq \mathfrak{p}\},$$

where the subset $S \subseteq R$ may as well be taken to be an *ideal* of R. It is clear that the set B_{zar} consisting of all (principal) open subsets of the form $X_R(f)$ for some $f \in R$, forms a basis for the Zariski topology.

Let us also point out the following straightforward result:

(2.3) **Lemma.** *An open subset U of $Spec(R)$ is quasicompact if and only if it is of the form $U = X(I)$ for some finitely generated R-ideal I.*

Proof. Assume U is quasicompact, then we may cover U by a *finite* number of $X(s_i) \in B_{zar}$. But then, with $I = \sum_i Rs_i$, we obtain $U = \bigcup_i X(s_i) = X(I)$, and I is finitely generated, indeed.

Conversely, assume that $U = X(I)$, with I finitely generated. If U is covered by a family of open subsets $\{X(I_\alpha); \alpha \in A\}$, then, from

$$X(I) = \bigcup_\alpha X(I_\alpha) = X(\sum_\alpha I_\alpha),$$

it follows that $I \subseteq rad(I) = rad(\sum_\alpha I_\alpha)$. Hence, since I is finitely generated, $I^n \subseteq \sum_\alpha I_\alpha$, for some positive integer n. But then, we may find a finite subset $B \subset A$, with $I^n \subseteq \sum_{\alpha \in B} I_\alpha$, and of course,

$$rad(I) = rad(I^n) \subseteq rad(\sum_{\alpha \in B} I_\alpha) \subseteq rad(\sum_\alpha I_\alpha),$$

whence equality, and

$$U = X(I) = X(\sum_{\alpha \in B} I_\alpha) = \bigcup_{\alpha \in B} X(I_\alpha),$$

which proves the assertion.   □

(2.4) If M is an R-module, then one may define a presheaf $\mathcal{Q}_{M,B}$ on B_{zar} by putting

$$\mathcal{Q}_{M,B}(X_R(f)) = M_f = R_f \otimes_R M,$$

the R-module of fractions of M with respect to the multiplicatively closed subset $\{1, f, f^2, \ldots\}$.

It is well known, cf. [38, 39, 88, 89, et al], that this defines a sheaf $\mathcal{O}_M$ on $Spec(R)$, with the following properties:

(2.4.1) for any $f \in R$, we have $\Gamma(X_R(f), \mathcal{O}_M) = M_f$;

(2.4.2) in particular, $\Gamma(Spec(R), \mathcal{O}_M) = M$;

(2.4.3) for any $\mathfrak{p} \in Spec(R)$, the stalk of $\mathcal{O}_M$ at $\mathfrak{p}$ is $\mathcal{O}_{M,\mathfrak{p}} = M_\mathfrak{p}$, the usual localization of M at $\mathfrak{p}$.

From this it easily follows that $\mathcal{O}_R$ is a sheaf of rings and that $\mathcal{O}_M$ is a sheaf of $\mathcal{O}_R$-modules for any $M \in R-\mathbf{mod}$.

Note also:

(2.4.4)   if $\varphi : R \to S$ is a ring homomorphism, then φ induces a morphism of ringed spaces

$$({}^a\varphi = f, \varphi_\#) : (Spec(S), \mathcal{O}_S) \to (Spec(R), \mathcal{O}_R);$$

the morphism $\varphi_\# : \mathcal{O}_R \to f_*\mathcal{O}_S$ yields φ by taking global sections;

(2.4.5)   with these notations, for any $M \in R-\mathbf{mod}$ and $N \in S-\mathbf{mod}$, we have $f_*\mathcal{O}_M = \mathcal{O}_{\varphi^*M}$ resp. $f^*\mathcal{O}_N = \mathcal{O}_{\varphi_*N}$, where φ^* resp. φ_* is the extension resp. restriction of scalars functor, i.e., $\varphi^*M = S \otimes_R M$ and $\varphi_*N = {}_RN$.

The role of localization theory comes into the present picture, if one wants to calculate sections of $\mathcal{O}_M$ over an arbitrary open subset $X_R(I)$ of $Spec(R)$.

To settle ideas, let us first assume R to be noetherian. In this case, for any ideal I of R, the radical σ_I is well defined and since $rad(I)^n \subseteq I$. for some positive integer n, it is clear that σ_I only depends upon the radical of I.

Let us start with the following observation:

(2.5) Lemma. *For any (commutative!) noetherian ring, any ideal I of R and any R-module M, there is a canonical isomorphism*

$$\varinjlim Hom_R(I^n, M) = \varinjlim Hom_R(I^n, M/\sigma_I M).$$

Proof. Since the fact that R is a commutative noetherian ring implies every radical over R to be stable, cf. (2.14), this follows immediately from (1.7) in the previous chapter. However, an easy, straightforward proof may be given as follows as well, cf. [89].

Let us denote by $\pi : M \to M' = M/\sigma_I M$ the canonical quotient map, then the morphism

$$\phi : \varinjlim Hom_R(I^n, M) = \varinjlim Hom_R(I^n, M/\sigma_I M)$$

is determined by sending any R-linear $f : I^n \to M$ to the composition $\pi f : I^n \to M'$. Using the fact that I is finitely generated, it is a straightforward exercise to verify that ϕ is injective.

To prove the surjectivity of ϕ, start from a morphism $f : I \to M'$. Since I is finitely generated, say by $\{x_1, \ldots, x_n\}$, there exists an R-linear, surjective morphism $p : R^n \to I$, which maps the canonical basic element $e_i \in R^n$ to $x_i \in I$. Denote the kernel of p by K, then there obviously exists a morphism $g : R^n \to M$ extending f, i.e., such that that the following diagram commutes:

$$
\begin{array}{ccc}
R^n & \xrightarrow{\ p\ } & I \\
{\scriptstyle g}\downarrow & & \downarrow{\scriptstyle f} \\
M & \xrightarrow[\ \pi\ ]{} & M'
\end{array}
$$

It follows that g maps K into $\sigma_I M$, hence $g(I^r K) = I^r g(K) = 0$, for some positive integer r, since $K \subseteq R^n$ is finitely generated, (as R is noetherian). This also implies, by Krull's Theorem, cf. [80, 99], that we may find some positive integer $s \geq r$, such that $I^s R^n \cap K \subseteq I^r K$. Since

$$Ker(p|I^s R^n) = I^s R^n \cap K \subseteq I^r K,$$

$g_1 = g|I^s R^n$ factorizes through $p(I^s R^n) = I^{s+1} \subseteq I$. So, there exists an R-linear map $f_1 : I^{s+1} \to M$, with the property that $f_1(p|I^s R^n) = g_1$, hence

$$\pi f_1(p|I^s R^n) = \pi g_1 = (f|I^{s+1})(p|I^s R^n),$$

i.e., $\pi f_1 = f|I^{s+1}$, since p is surjective. Denoting classes in the above inductive limits by $[-]$, we thus showed that $\phi([f_1]) = [f]$, which proves the surjectivity of ϕ, indeed. $\square$

From this, we may deduce the following result:

(2.6) Proposition. *(Deligne's formula) Let R be a noetherian ring and M an R-module. Then, for any ideal I of R, there is a canonical isomorphism*

$$\Gamma(X_R(I), \mathcal{O}_M) = \varinjlim Hom_R(I^n, M).$$

Proof. The ideal I being finitely generated, say by $\{s_1, \ldots, s_k\}$, we may cover $X_R(I)$ by the open subsets $X_R(s_i) \in B_{zar}$, and obtain an exact sequence

$$0 \to \Gamma(X_R(I), \mathcal{O}_M) \xrightarrow{\alpha} \bigoplus_i M_{s_i} \xrightarrow{\beta} \bigoplus_{i,j} M_{s_i s_j},$$

with obvious morphisms α and β.

Define a map

$$\mu : \varinjlim Hom_R(I^n, M) \to \Gamma(X_R(I), \mathcal{O}_M),$$

by sending a class $[f : I^n \to M]$ in the inductive limit to the corresponding family $(f_i(1)) \in Ker(\beta) = \Gamma(X_R(I), \mathcal{O}_M)$. Of course, here, f_i denotes the localized map

$$f_i = f_{s_i} : R_{s_i} = (I^n)_{s_i} \to M_{s_i}.$$

Conversely, define by induction a map

$$\nu : \Gamma(X_R(I), \mathcal{O}_M) \to \varinjlim Hom_R(I^n, M),$$

as follows. Choose $(\alpha_i = a_i/s_i^n) \in \Gamma(X_R(I), \mathcal{O}_M) = Ker(\beta)$. Up to replacing s_i by s_i^n, we may assume $n = 1$. Let $L_1 = Rs_1$ and define a morphism $f_1 : L_1 \to M/\sigma_1 M$, by sending $rs_1 \in L_1$ to $M/\sigma_1 M$, where, for each i, we denote by σ_i the radical associated to the ideal Rs_i of R. By the previous Lemma, we may find some positive integer n and an R-linear map $g_1 : Rs_1^n = L_1^n \to M$, such that the composition

$$L_1^n \xrightarrow{g_1} M \xrightarrow{p_1} M/\sigma_1 M$$

is just the restriction $f_1|L_1^n$. Let $b_1 = g_1(s_1^n)$, then, up to replacing a_1 by b_1 and s_1 by s_1^n, this yields a morphism $h_1 : Rs_1 \to M$ such that $\alpha_1 = a_1/s_1$, with $a_1 = h_1(s_1)$. Moreover, without any loss of generality, we may even assume that $a_i s_j = a_j s_i$, for any pair of indices i, j.

Let us now argue inductively. Suppose, for some $t < k$, that we have defined a morphism

$$h_t : L_t = \sum_{i \leq t} Rs_i \longrightarrow M,$$

and succeeded in representing α_i as a_i/s_i, with $a_i = h_t(s_i)$, for any $i \leq t$ and with $a_i s_j = a_j s_i$, for any pair of indices i, j. We then want to define $h_{t+1} : L_{t+1} \to M$, with similar properties. One realizes this, by first constructing $f_{t+1} : Rs_{t+1} \to M/\sigma_{t+1}M$, exactly as we did for f_1. We thus find representations a_i/s_i of the α_i, with $a_i s_j = a_j s_i$, for each pair of indices i, j, together with $h_t : L_t \to M$ and $f_{t+1} : Rs_{t+1} \to M$, such that $h_t(s_i) = a_i$, for $1 \leq i \leq t$ and $f_{t+1}(s_{t+1}) = a_{t+1}$. Let $L_{t+1} = L_t + Rs_{t+1}$, and denote the associated quotient map by $p_{t+1} : M \to M/\sigma_{t+1}M$. It is easy to verify that this yields a well-defined map $l_{t+1} : L_{t+1} \to M/\sigma_{t+1}M$, by letting $l_{t+1}|L_t = p_{t+1}h_t$ and $l_{t+1}|Rs_{t+1} = p_{t+1}f_{t+1}$. The foregoing Lemma then allows us to "lift" l_{t+1} to some $g_{t+1} : L_{t+1}^n \to M$, with the same nice features as the previously constructed h_i.

By induction, we thus obtain a morphism $h_k : L_k \to M$, such that L_k is generated by the elements $s_1, \ldots, s_k$, with the property that $h_k(s_i)/s_i = \alpha_i$, for all i. Since these s_i, generating L_k, have been obtained by taking successive powers of the initial s_i, clearly $rad(L_k) = rad(I)$, hence $I^k \subseteq L_k$, for some positive integer n. The map ν, announced above, is now defined by sending the family $(\alpha_i; i = 1, \ldots, k)$ to the class $[h_k|I^n : I^n \to M]$. We leave it as a straightforward exercise, to check that μ and ν are inverse to each other. □

(2.7) Example. As a typical example, consider the polynomial ring $R = \mathbf{C}[x, y]$ with associated spectrum $Spec(R) = \mathbf{A}_{\mathbf{C}}^2$. If $\mathfrak{m}$ denotes the (maximal) ideal $Rx + Ry = (x, y)$ of R, then obviously $U = X_R(\mathfrak{m}) = A_{\mathbf{C}}^2 \setminus \{(0, 0)\}$. As $U = X_R(x) \cup X_R(y)$, it is easy to calculate the sections of $\mathcal{O}_R$ over U. Indeed, since

$$\Gamma(X_R(x), \mathcal{O}_R) = R_x \subseteq \mathbf{C}(x, y)$$
$$\Gamma(X_R(y), \mathcal{O}_R) = R_y \subseteq \mathbf{C}(x, y)$$

we find that

$$\Gamma(U, \mathcal{O}_R) = R_x \cap R_y = R.$$

Alternatively, applying Deligne's formula, we also find directly

$$\Gamma(U = X_R(\mathfrak{m}), \mathcal{O}_R) = \{z \in \mathcal{C}(x, y); \ \exists n \in N, \mathfrak{m}^n z \subseteq R = \mathbf{C}[x, y]\}$$
$$= R,$$

the last equality, essentially because R is integrally closed.

Let us now show, how a little torsion theory not only allows us to simplify, but even to generalize the foregoing result.

(2.8) Let R be an arbitrary commutative ring, not necessarily noetherian anymore, and let Y be a subset of $Spec(R)$. We may associate to Y the set $\mathcal{L}(Y)$, which consists of all ideals I of R, with the property that $Y \subseteq X_R(I)$. We leave it as a straightforward exercise to the reader, to verify that $\mathcal{L}(Y)$ is actually a Gabriel filter. We denote by σ_Y the associated radical in $R-\mathbf{mod}$. Since obviously $\mathcal{L}(Y)$ and σ_Y only depend upon the generic closure of Y, we may as well assume Y to be generically closed from the start.

Conversely, if $\mathcal{L}$ is a filter of ideals of R, then we may associate to it a (generically closed) subset $Y(\mathcal{L}) = Spec(R) \setminus \mathcal{L}$ of $Spec(R)$. Of course, if $\mathcal{L} = \mathcal{L}(\sigma)$, then $Y(\mathcal{L}) = \mathcal{K}(\sigma)$.

Clearly, if $Y \subseteq Spec(R)$ is generically closed, then $Y(\mathcal{L}(Y)) = Y$. Indeed, $\mathfrak{p} \in Y(\mathcal{L}(Y))$ is equivalent to $Y \not\subseteq X_R(\mathfrak{p})$, i.e., we may find some $\mathfrak{q} \in Y$ with $\mathfrak{q} \notin X_R(\mathfrak{p})$, or, equivalently, with $\mathfrak{p} \subseteq \mathfrak{q}$. Since this is equivalent to $\mathfrak{p} \in Y$, as Y is generically closed, this proves the assumption. Conversely, if $\mathcal{L}$ is a Gabriel filter in $R-\mathbf{mod}$, then, by definition, $\mathcal{L}(Y(\mathcal{L})) = \mathcal{L}$ if and only if for every $I \notin \mathcal{L}$, we may find some prime ideal $\mathfrak{p} \supseteq I$, with $\mathfrak{p} \notin \mathcal{L}$.

(2.9) Lemma. *Assume the Gabriel filter $\mathcal{L}$ on R is of finite type. Then $\mathcal{L}$ is of the form $\mathcal{L}(Y)$, for some (generically closed) subset $Y \subseteq Spec(R)$.*

Proof. If $I \notin \mathcal{L}$, then we may apply Zorn's Lemma in order to find an ideal $\mathfrak{p} \supseteq I$, which is maximal with respect to the property of not being contained in $\mathcal{L}$.

Indeed, for any chain $I \subseteq I_1 \subseteq I_2 \subseteq \ldots$ of ideals not contained in $\mathcal{L}$, we have $J = \bigcup_n I_n \notin \mathcal{L}$. For, otherwise, pick a finitely generated $K \in \mathcal{L}$ with $K \subseteq J$. Since K then has to be contained in some I_n, this implies $I_n \in \mathcal{L}$ as well – a contradiction.

Moreover, it is straightforward to see that $\mathfrak{p}$ is actually a prime ideal, in this case. From the remarks preceding the Lemma, it thus trivially follows that $\mathcal{L} = \mathcal{L}(Y(\mathcal{L}))$. $\square$

In particular, every radical of finite type σ in $R-\mathbf{mod}$ is of the form $\sigma = \sigma_Y$ for some generically closed subset $Y \subseteq Spec(R)$.

(2.10) Corollary. *If the radical σ in $R-\mathbf{mod}$ has finite type, then $\mathcal{L}(\sigma) = \mathcal{L}(\mathcal{K}(\sigma))$. In particular, σ and $\mathcal{K}(\sigma)$ completely determine each other.*

Of course, under the assumptions of the previous Corollary, an R-module M belongs to $\mathcal{T}_\sigma$ if and only if $M_\mathfrak{p} = 0$, for all $\mathfrak{p} \in \mathcal{K}(\sigma)$. In particular, an R-linear map $u : M \to N$ is a σ-isomorphism, if and only if $u_\mathfrak{p} : M_\mathfrak{p} \to N_\mathfrak{p}$ is an isomorphism, for all $\mathfrak{p} \in \mathcal{K}(\sigma)$.

(2.11) Let us now assume I to be a finitely generated ideal of R. As pointed out before, this implies, in particular, the associated open set

$X_R(I) \subseteq Spec(R)$ to be quasicompact. Moreover, since I is finitely generated, we know that this yields a radical σ_I of finite type in $R-\mathbf{mod}$, defined by putting for any R-module M

$$\sigma_I M = \{m \in M; \; \exists n \in N, I^n m = 0\}.$$

If we again denote by $Q_I(-)$ the associated radical in $R-\mathbf{mod}$, then we obtain the following generalization of Deligne's formula:

(2.12) Proposition. *Let I be a finitely generated ideal of R. Then for any R-module M there is a canonical isomorphism*

$$\Gamma(X_R(I), \mathcal{O}_M) = Q_I(M).$$

Proof. Since I is finitely generated, $X_R(I)$ is quasicompact, and we may cover it with a finite family of open sets $\{X_R(s_i); i = 1, \ldots, n\} \subseteq B_{zar}$. Let us write M_i, etc. for the localization of M at s_i, etc. Since $\mathcal{O}_M$ is a sheaf, we obtain an exact sequence

$$0 \to \Gamma(X_R(I), \mathcal{O}_M) \to \bigoplus_i M_{s_i} \to \bigoplus_{i,j} M_{s_i s_j} \quad (*),$$

with obvious morphisms. For any $1 \leq k \leq n$, localizing at s_k yields an exact sequence

$$0 \to \Gamma(X_R(I), \mathcal{O}_M)_{s_k} \to \bigoplus_i M_{s_i s_k} \to \bigoplus_{i,j} M_{s_i s_j s_k} \quad (**).$$

On the other hand, the open covering $\{X_R(s_i s_k); 1 \leq i \leq n\}$ of $X_R(s_k)$, yields an exact sequence

$$0 \to M_{s_k} \to \bigoplus_i M_{s_i s_k} \to \bigoplus_{i,j} M_{s_i s_j s_k},$$

where the map $\bigoplus_i M_{s_i s_k} \to \bigoplus_{i,j} M_{s_i s_j s_k}$ is the same one as in $(**)$. It follows that the canonical morphism $u : M \to \Gamma(X_{(}I), \mathcal{O}_M)$ yields for each index k an isomorphism $u_{s_k} : M_{s_k} \to \Gamma(X_R(I), \mathcal{O}_M)_{s_k}$, hence, for each $\mathfrak{p} \in X_R(s_k) \subseteq X_R(I)$, by localizing, an isomorphism $u_{\mathfrak{p}}$:

$M_{\mathfrak{p}} \to \Gamma(X_R(I), \mathcal{O}_M)_{\mathfrak{p}}$. To finish the proof, it thus suffices to verify that $\Gamma(X_R(I), \mathcal{O}_M)$ is σ_I-closed, but this is an immediate consequence of the exactness of (*) and the fact that each M_{s_i} is σ_I-closed, since $\sigma_I \leq \sigma_{s_i}$.   □

(2.13) Lemma. Let σ and τ be radicals of finite type in $R-\mathbf{mod}$. Then for any R-module M, we have $Q_\sigma(\tau M) = \tau Q_\sigma(M)$.

Proof. Let $m \in Q_\sigma(\tau M)$, then $Im \subseteq \tau M/\sigma(\tau M)$, for some finitely generated $I \in \mathcal{L}(\sigma)$. Since for each $i \in I$, we may pick $m_i \in \tau M$ with $im = \overline{m_i} = m_i \bmod \sigma(\tau M)$, we may find for any such i an ideal $J_i \in \mathcal{L}(\tau)$, with $J_i im = J_i \overline{m_i} = 0$, hence, as I is finitely generated, a single $J \in \mathcal{L}(\tau)$, such that $JIm = 0$. It follows that $Jm \subseteq \sigma Q_\sigma(\tau M) = 0$, hence that $m \in \tau Q_\sigma(M)$.

Conversely, if $m \in \tau Q_\sigma(M)$, then there exist finitely generated ideals $I \in \mathcal{L}(\sigma)$ and $J \in \mathcal{L}(\tau)$, with the property that $Im \subseteq M/\sigma M$ and $Jm = 0$. For any $i \in I$, we may find some $m_i \in M$ such that $im = \overline{m_i} = m_i \bmod \sigma M$. Since $Jm = 0$, we thus find that $Jm_i \subseteq \sigma M$, and as J is finitely generated, there exists some $I'_i \in \mathcal{L}(\sigma)$, with $JI'_i m_i = I'_i Jm_i = 0$, hence $I'_i m_i \subseteq \tau M$. Finally, it follows that $I'_i \overline{m_i} \subseteq \tau M/\sigma(\tau M)$, hence $I'_i im \subseteq \tau M/\sigma(\tau M)$ and $im \in Q_\sigma(\tau M)$. So, $Im \subseteq Q_\sigma(\tau M)$, which proves the assertion.   □

For noetherian rings, this now immediately yields:

(2.14) Corollary. Assume R to be noetherian and let M be an R-module. Then, for any $\mathfrak{p} \in Spec(R)$ and any radical $\sigma \in R-\mathbf{mod}$, we have $Q_\sigma(M_{\mathfrak{p}}) = Q_\sigma(M)_{\mathfrak{p}}$.

Proof. Since both σ and $\sigma_{R\backslash\mathfrak{p}}$ have finite type, the previous result easily implies that $j_{\sigma,M} : M \to Q_\sigma(M)$ induces a σ-isomorphism $M_{\mathfrak{p}} \to Q_\sigma(M)_{\mathfrak{p}}$. The localization morphism $j_{\sigma,M_{\mathfrak{p}}} : M_{\mathfrak{p}} \to Q_\sigma(M_{\mathfrak{p}})$ thus extends to a linear map $h : Q_\sigma(M)_{\mathfrak{p}} \to Q_\sigma(M_{\mathfrak{p}})$. Since the localization functors at any pair of prime ideals of R commute, clearly for each

$\mathfrak{q} \in \mathcal{K}(\sigma)$ the localized map $h_\mathfrak{q}$ reduces to the identity on

$$(Q_\sigma(M)_\mathfrak{p})_\mathfrak{q} = Q_\sigma(M_\mathfrak{p})_\mathfrak{q} = (M_\mathfrak{p})_\mathfrak{q},$$

i.e., h is a σ-isomorphism.

To prove the assertion, it thus suffices to show that $Q_\sigma(M)_\mathfrak{p}$ is σ-closed. However, this follows easily from the fact that for each ideal I of $\mathcal{L}(\sigma)$ (in particular, for $I = R!$)

$$Hom_R(I, Q_\sigma(M)_\mathfrak{p}) = Hom_R(I_\mathfrak{p}, Q_\sigma(M)_\mathfrak{p}) = Hom_R(I, Q_\sigma(M))_\mathfrak{p},$$

as R is assumed to be noetherian.   □

Somewhat more generally, using sheaf theoretic methods, we may also prove:

(2.15) Lemma. *Let σ be a radical of finite type in $R-\mathbf{mod}$ and let $\mathfrak{p}$ be a prime ideal of R. Then for any R-module M, we have a canonical isomorphism*

$$Q_\sigma(M)_\mathfrak{p} = Q_\sigma(M_\mathfrak{p}).$$

Proof. Let us first assume I to be a finitely generated ideal of R, then we may write $X_R(I) = \bigcup_i X_R(f_i)$, for some finite number of elements f_i of R. Taking sections of $\mathcal{O}_{M_\mathfrak{p}}$ over $X_R(I)$ yields an exact sequence

$$0 \to Q_I(M_\mathfrak{p}) \to \bigoplus_i (M_\mathfrak{p})_{f_i} \to \bigoplus_{i,j} (M_\mathfrak{p})_{f_i f_j}.$$

Arguing similarly for M and localizing at the prime $\mathfrak{p}$, yields an exact sequence

$$0 \to Q_I(M)_\mathfrak{p} \to \bigoplus_i (M_{f_i})_\mathfrak{p} \to \bigoplus_{i,j} (M_{f_i f_j})_\mathfrak{p}.$$

Since the last two terms in the previous exact sequences are obviously isomorphic, it follows that $Q_I(M)_\mathfrak{p} = Q_I(M_\mathfrak{p})$. This proves the assertion, for $\sigma = \sigma_I$.

In the general case, first note that $M_\mathfrak{p}/\sigma M_\mathfrak{p} = M_\mathfrak{p}/(\sigma M)_\mathfrak{p} = (M/\sigma M)_\mathfrak{p}$, hence, to prove that $Q_\sigma(M)_\mathfrak{p} = Q_\sigma(M_\mathfrak{p})$, we may assume that M is σ-torsionfree. But then, clearly, all of the maps $Q_I(M) \to Q_\sigma(M)$, induced by $\sigma_I \leq \sigma$, for $I \in \mathcal{L}(\sigma)$, are injective. We thus find that $Q_\sigma(M) = \bigcup Q_I(M)$, where I runs through (the cofinal subset of) all finitely generated ideals $I \in \mathcal{L}(\sigma)$. So,

$$Q_\sigma(M)_\mathfrak{p} = \bigcup Q_I(M)_\mathfrak{p} = \bigcup Q_I(M_\mathfrak{p}) = Q_\sigma(M_\mathfrak{p}),$$

and this proves the assertion.   □

(2.16) Example. If σ does not have finite type, then the foregoing result is not necessarily true anymore. Indeed, let us consider the following example, essentially due to F. Call [22].

As in (3.19) in Chapter I, let k be an arbitrary field and let $R \subseteq k^N$ be the ring consisting of all sequences that eventually become constant. The ring R has a non-finitely generated maximal ideal $\mathfrak{m} = \bigoplus_i Re_i$, where e_i is the sequence that has 1 at the ith place and zero elsewhere. Alternatively, $\mathfrak{m}$ consists of all sequences that eventually become zero. Since $\mathfrak{m} = \mathfrak{m}^2$, it defines a radical $\sigma_\mathfrak{m}$ in $R-\mathbf{mod}$ with Gabriel filter $\mathcal{L}(\sigma_\mathfrak{m}) = \{\mathfrak{m}, R\}$ and torsion class $\mathcal{T}_\mathfrak{m}$ consisting of all $M \in R-\mathbf{mod}$ with $\mathfrak{m}M = 0$.

As pointed out before, R is torsionfree at $\sigma_\mathfrak{m}$, so $R \subseteq Q_\mathfrak{m}(R)$, where the localization $Q_\mathfrak{m}(R)$ of R at $\sigma_\mathfrak{m}$ is easily seen to be actually isomorphic to k^N. Since $R_\mathfrak{m}$ (the usual localization of R at the maximal ideal $\mathfrak{m}$) is nonzero, as R is certainly not torsion at $\mathfrak{m}$, we thus find that $Q_\mathfrak{m}(R)_\mathfrak{m} \neq 0$, as well.

On the other hand, it is easy to see that $\mathfrak{m}$ is $\sigma_{R\backslash\mathfrak{m}}$-torsion. Indeed, if $m \in \mathfrak{m}$ is zero from position i on, then $sm = 0$, where s is zero up to position $i - 1$ and 1 afterwards. As $s \notin \mathfrak{m}$, it follows that $m \in \sigma_{R\backslash\mathfrak{m}}\mathfrak{m}$, which proves our claim. It thus follows that $\mathfrak{m}R_\mathfrak{m} = \mathfrak{m}_\mathfrak{m} = 0$, showing that $R_\mathfrak{m} \in \mathcal{T}_\mathfrak{m}$. So, $Q_\mathfrak{m}(R_\mathfrak{m}) = 0$. We thus find that $Q_\mathfrak{m}(R_\mathfrak{m}) \neq Q_\mathfrak{m}(R)_\mathfrak{m}$, which yields the desired counterexample.

(2.17) Corollary. *If σ and τ are radicals of finite type in $R-\mathbf{mod}$, then $Q_\sigma Q_\tau = Q_\tau Q_\sigma$.*

Proof. If $M \in R-\mathbf{mod}$, then the canonical map

$$Q_\tau(j_{\sigma,M}) : Q_\tau(M) \to Q_\tau(Q_\sigma(M))$$

is a σ-isomorphism, as one easily verifies by localizing at any $\mathfrak{q} \in \mathcal{K}(\sigma)$ and applying the previous Lemma. Hence the localization map $Q_\tau(M) \to Q_\sigma(Q_\tau(M))$ extends to an R-linear map

$$u : Q_\tau(Q_\sigma(M)) \to Q_\sigma(Q_\tau(M)).$$

One obtains in a similar way an R-linear map

$$v : Q_\sigma(Q_\tau(M)) \to Q_\tau(Q_\sigma(M)),$$

which, by an obvious unicity argument, is easily seen to be the inverse of u. $\square$

(2.18) Note. In the noetherian case, the previous result also follows from (3.24) below, taking into account the fact that all radicals in a noetherian commutative ring are stable. The direct proof we gave avoids torsion-theoretic technicalities, however.

(2.19) Lemma. *If σ and τ are radicals of finite type in $R-\mathbf{mod}$, then $\mathcal{L}(\mathcal{K}(\sigma) \cap \mathcal{K}(\tau))$ consists of all ideals L of R, which contain a product IJ, where $I \in \mathcal{L}(\sigma)$ and $J \in \mathcal{L}(\tau)$.*

Proof. It is fairly easy to see that the set $\mathcal{L}$ consisting of all ideals L of R containing a product IJ with $I \in \mathcal{L}(\sigma)$ and $J \in \mathcal{L}(\tau)$ is a Gabriel filter of finite type. Now, for any prime ideal $\mathfrak{p}$ of R, we have $\mathfrak{p} \in \mathcal{L}$ if and only if $I \not\subset \mathfrak{p}$ for all $I \in \mathcal{K}(\sigma) \cap \mathcal{K}(\tau)$, i.e., if $\mathfrak{p} \in \mathcal{K}(\sigma) \cap \mathcal{K}(\tau)$. It thus follows that $\mathcal{L} = \mathcal{L}(\mathcal{K}(\sigma) \cap \mathcal{K}(\tau))$, indeed. $\square$

(2.20) Corollary. *Let σ and τ be radicals of finite type in $R-\mathbf{mod}$. Then $Q_{\sigma \vee \tau} = Q_\sigma Q_\tau$.*

Proof. For any R-module M, we claim that $Q_\sigma Q_\tau(M) = Q_\tau Q_\sigma(M)$ is $\sigma \vee \tau$-closed.

Indeed, it is easy to verify that $Q_\sigma Q_\tau(M)$ is $\sigma \vee \tau$-torsionfree, using the above characterization of $\mathcal{L}(\sigma \vee \tau)$.

On the other hand, pick some $I \in \mathcal{L}(\sigma)$ and some $J \in \mathcal{L}(\tau)$, then any morphism $\phi : IJ \to Q_\sigma Q_\tau(M)$ extends to $\phi' : J \to Q_\sigma Q_\tau(M)$ (since J/IJ is σ-torsion and $Q_\sigma Q_\tau(M)$ is σ-closed), and hence to some $\phi'' : R \to Q_\sigma Q_\tau(M)$ (since R/J is τ-torsion, and $Q_\sigma Q_\tau(M)$ is also τ-closed). More generally, pick $L \in \mathcal{L}(\sigma \vee \tau)$, and consider a morphism $\psi : L \to Q_\sigma Q_\tau(M)$. To prove that ψ extends to R, first restrict it to some IJ, with $I \in \mathcal{L}(\sigma)$ and $J \in \mathcal{L}(\tau)$, then the induced map $\phi : IJ \to Q_\sigma Q_\tau(M)$ extends to $\phi'' : R \to Q_\sigma Q_\tau(M)$, as previously. But $\phi''|L = \psi$. Indeed, if $\lambda = \phi''|L - \psi$, then $\lambda|IJ = 0$, hence λ factorizes through L/IJ by some map $\lambda' : IJ/L \to Q_\sigma Q_\tau(M)$. But, L/IJ is $\sigma \vee \tau$-torsion. i.e., $\lambda' = 0$, which proves the assertion.

Finally, consider the composition

$$j : M \xrightarrow{u} Q_\tau(M) \xrightarrow{v} Q_\sigma Q_\tau(M).$$

For any $\mathfrak{p} \in \mathcal{K}(\sigma)$, the map $v_{\mathfrak{p}}$ is an isomorphism, so the localized map $j_{\mathfrak{p}} : M_{\mathfrak{p}} \to Q_\sigma Q_\tau(M)_{\mathfrak{p}}$ is just the morphism $u_{\mathfrak{p}} : M_{\mathfrak{p}} \to Q_\tau(M)_{\mathfrak{p}}$. So, if $p \in \mathcal{K}(\sigma) \cap \mathcal{K}(\tau)$, then $j_{\mathfrak{p}}$ is an isomorphism, since $u_{\mathfrak{p}}$ then also is. It follows that $Ker(j)$ and $Coker(j)$ are both $\sigma \vee \tau$-torsion, i.e., j induces an isomorphism

$$Q_{\sigma \vee \tau}(M) = Q_{\sigma \vee \tau} Q_\sigma Q_\tau(M) = Q_\sigma Q_\tau(M),$$

and this finishes the proof. $\square$

As we have seen in the previous Section, any continuous map $f : X \to Y$ between topological spaces induces a functor $f_p^{-1} : \mathcal{P}(Y) \to \mathcal{P}(X)$ and a functor $f^{-1} : \mathcal{S}(Y) \to \mathcal{S}(X)$. If X is a subset of Y with the induced topology and if f is the inclusion $X \hookrightarrow Y$, then, for any $\mathcal{F} \in \mathcal{S}(Y)$, we write $\mathcal{F}|X$ for $f^{-1}\mathcal{F}$. We also write $\Gamma(Y, \mathcal{F})$ for $\Gamma(Y, \mathcal{F}|Y)$.

(2.21) Proposition. *Let σ be a radical of finite type in $R-\mathbf{mod}$ and let M be an R-module. Then there is a canonical isomorphism*

$$\Gamma(\mathcal{K}(\sigma), \mathcal{O}_M) = Q_\sigma(M).$$

Proof. Let $f : Y = \mathcal{K}(\sigma) \hookrightarrow Spec(R)$ denote the canonical inclusion and write $Y(I) = X_R(I) \cap Y = f^{-1}(X_R(I))$ for any ideal I of R. Since we assumed the topology on Y to be induced from $Spec(R)$, every open subset of Y is of this form. Hence the $Y(I)$ with I finitely generated form a basis B for the topology on Y. Now, the presheaf $f_p^{-1}\mathcal{O}_M$ on Y is given by

$$(f_p^{-1}\mathcal{O}_M)(Y(I)) = \varinjlim \mathcal{O}_M(X_R(J)),$$

where $X_R(J) \supseteq Y(I)$, i.e., with $J \in \mathcal{L}(Y(I))$. If we choose the ideal I to be finitely generated, then we may clearly restrict this inductive limit to ideals J, which are finitely generated as well. So,

$$(f_p^{-1}\mathcal{O}_M)(Y(I)) = \varinjlim Q_J(M),$$

where J runs through the finitely generated ideals in $\mathcal{L}(Y(I))$. For each such J there is an obvious map $Q_J(M) \to Q_{Y(I)}(M)$, hence a morphism

$$t_I : \varinjlim Q_J(M) \longrightarrow Q_{Y(I)}(M).$$

We may define a presheaf of $\mathcal{O}_M|Y$-modules $\mathcal{H}_M$ on B, by putting $\mathcal{H}(Y(I)) = Q_{Y(I)}(M)$, for each finitely generated ideal I of R.

From the preceding results, it now follows for any $\mathfrak{p} \in Y$, that the stalk of $\mathcal{H}_M$ at $\mathfrak{p}$ is given by

$$\mathcal{H}_{M,\mathfrak{p}} = \varinjlim_{\mathfrak{p} \in Y(I)} Q_{Y(I)}(M) = \varinjlim_{\mathfrak{p} \in Y(I)} Q_I Q_\sigma(M) = Q_\sigma(M)_\mathfrak{p} = M_\mathfrak{p}.$$

It thus follows that the t_I yield a presheaf morphism

$$t : f_p^{-1}\mathcal{O}_M \longrightarrow \mathcal{H}_M$$

on B, which is an isomorphism at all $\mathfrak{p} \in Y$.

In order to finish the proof, it is now clearly sufficient to show that $\mathcal{H}_M$ is a sheaf on B. Indeed, the morphism t then induces a canonical isomorphism $f^{-1}\mathcal{O}_M \cong \mathcal{H}_M$, and we then obtain that

$$\Gamma(Y, \mathcal{O}_M) = \Gamma(Y, f^{-1}\mathcal{O}_M) = \Gamma(Y, \mathcal{H}_M) = Q_\sigma(M).$$

Choose an open covering $\{Y(I_\alpha); \alpha \in A\}$ in B of some $Y(I)$. If $m \in \mathcal{H}_M(Y(I))$ is mapped to 0 by all restriction maps $\mathcal{H}_M(Y(I)) \to \mathcal{H}_M(Y(I_\alpha))$, then by definition $m \in \sigma_{Y(I)}Q_{Y(I)}(M) = 0$. This shows that the presheaf $\mathcal{H}_M$ is certainly separated.

On the other hand, assume that for all $\alpha \in A$, there is given some $m_\alpha \in \mathcal{H}_M(Y(I_\alpha)) = Q_{Y(I_\alpha)}(M)$, such that

$$m_\alpha | Y(I_\alpha) \cap Y(I_\beta) = m_\beta | Y(I_\alpha) \cap Y(I_\beta),$$

for any $\alpha, \beta \in A$. As $Y(I) = \bigcup_{\alpha \in A} Y(I_\alpha)$, clearly $Y(I) \subseteq X_R(\sum_{\alpha \in A} I_\alpha)$, i.e., $\sum_{\alpha \in A} \in \mathcal{L}(Y(I))$, by definition. But, as $\mathcal{L}(Y(I)) = \mathcal{L}(\mathcal{K}(\sigma) \cap X_R(I))$, it follows from (2.19) that for some finitely generated $L \in \mathcal{L}(Y(I))$, we have $L \subseteq \sum_{\alpha \in A} I_\alpha$. We may thus find a finite number of ideals $I_1, \ldots, I_n$ amongst the I_α such that $K = \sum_{i=1}^n I_i \in \mathcal{L}(Y(I))$. Hence $Y(I) \subseteq X(K) = X(I_1) \cup \ldots \cup X(I_n)$, so $Y(I)$ is covered by $\{Y(I_1), \ldots, Y(I_n)\}$. Since $\mathcal{O}_M$ is a sheaf on $Spec(R)$, there is an exact sequence

$$0 \longrightarrow \Gamma(X(K), \mathcal{O}_M) \longrightarrow \bigoplus_i \Gamma(X(I_i), \mathcal{O}_M) \Longrightarrow \bigoplus_{i,j} \Gamma(X(I_iI_j), \mathcal{O}_M)$$

with obvious morphisms. This diagram reduces to

$$0 \longrightarrow Q_K(M) \longrightarrow \bigoplus_i Q_{I_i}(M) \Longrightarrow \bigoplus_{i,j} Q_{I_iI_j}(M).$$

Localizing at σ yields an exact sequence

$$0 \longrightarrow Q_\sigma Q_K(M) \longrightarrow \bigoplus_i Q_\sigma Q_{I_i}(M) \Longrightarrow \bigoplus_{i,j} Q_\sigma Q_{I_iI_j}(M).$$

Now, $Q_\sigma Q_K(M) = Q_{Y(K)}(M) = Q_{Y(I)}(M)$, and similarly for the other localizations, so the previous exact sequence reduces to

$$0 \longrightarrow \Gamma(Y(K), \mathcal{H}_M) \longrightarrow \bigoplus_i \Gamma(Y(I_i), \mathcal{H}_M) \Longrightarrow \bigoplus_{i,j} \Gamma(Y(I_iI_j), \mathcal{H}_M).$$

We thus find some $m \in \mathcal{H}_M(Y(I))$ with $m|Y(I_i) = m_i$, for $1 \leq i \leq n$. To conclude, for any $\alpha \in A$ and any $1 \leq i \leq n$, we have

$$
\begin{aligned}
(m|Y(I_\alpha))|Y(I_\alpha) \cap Y(I_i) &= (m|Y(I_i))|Y(I_\alpha) \cap Y(I_i) \\
&= m_i|Y(I_i) \cap Y(I_\alpha) \\
&= m_\alpha|Y(I_i) \cap Y(I_\alpha).
\end{aligned}
$$

Hence $m|Y(I_\alpha) = m_\alpha$, since $Y(I_\alpha)$ is covered by the $Y(I_\alpha) \cap Y(I_i)$. This shows that $\mathcal{H}_M$ is a sheaf and finishes the proof. $\qquad \square$

3. Compatibility in module categories

(3.1) Throughout, R will denote an arbitrary, not necessarily noetherian ring.

Let σ and τ denote radicals in $R-\mathbf{mod}$. Following [84, 90], we will say that σ is Q_τ-*compatible* if the functors σ and Q_τ commute, i.e., if $\sigma Q_\tau(M) = Q_\tau(\sigma M)$, for any $M \in R-\mathbf{mod}$. We will say that σ and τ are *mutually compatible* if $\sigma Q_\tau = Q_\tau \sigma$ and $\tau Q_\sigma = Q_\sigma \tau$, i.e., if σ is Q_τ-compatible and τ is Q_σ-compatible. For example, it follows from (2.13) that *over a commutative ring R*, any pair of radicals of finite type are mutually compatible. We will also see in (3.8) below that, somewhat surprisingly, these conditions are equivalent, i.e., we will prove that $\sigma Q_\tau = Q_\tau \sigma$ implies that $\tau Q_\sigma = Q_\sigma \tau$.

(3.2) Example. If $\sigma \geq \tau$, then σ is Q_τ-compatible.

Proof. Let M be a left R-module and pick $m \in Q_\tau(\sigma M)$, then we may find some $I \in \mathcal{L}(\tau)$, with the property that $Im \subseteq \sigma M/\tau M$. As $(Ann_R^\ell(m) : r)$ belongs to $\mathcal{L}(\sigma)$ for every $r \in I$ and as $I \in \mathcal{L}(\tau) \subseteq \mathcal{L}(\sigma)$, it follows that $Ann_R^\ell(m) \in \mathcal{L}(\sigma)$. So, $m \in \sigma Q_\tau(M)$, i.e., $Q_\tau(\sigma M) \subseteq \sigma Q_\tau(M)$.

Conversely, it is obvious that $M/\sigma M \in \mathcal{F}_\tau$, so this yields an essential monomorphism $M/\sigma M \subseteq Q_\tau(M/\sigma M)$. Since

$$\sigma(Q_\tau(M)/Q_\tau(\sigma M)) \cap M/\sigma M = 0,$$

within $Q_\tau(M/\sigma M)$, it follows that $\sigma(Q_\tau(M)/Q_\tau(\sigma M)) = 0$. But then, $\sigma Q_\tau(M) = \sigma Q_\tau(\sigma M) \subseteq Q_\tau(\sigma M)$, which yields the other inclusion.

$\square$

Clearly, if $\sigma \geq \tau$, then $Q_\sigma Q_\tau(M) = Q_\tau(M)$, for any $M \in R-\mathbf{mod}$. It has been proved in [84, 87], that this result generalizes as follows:

(3.3) Proposition. *Let R be a ring, let σ and τ be radicals in $R-\mathbf{mod}$ and denote by $\overline{\sigma}$ the radical in $Q_\tau(R)-\mathbf{mod}$ induced by σ through the* ring homomorphism $j_{\tau,R} : R \to Q_\tau(R)$. If σ is Q_τ-compatible, then for every left R-module M, we have

$$Q_{\overline{\sigma}}Q_\tau(M) = Q_\sigma Q_\tau(M).$$

Proof. Let M be a left R-module, then, since σ is Q_τ compatible, it is easy to see that

$$\overline{\sigma}Q_\tau(M) = \sigma Q_\tau(M) = Q_\tau(\sigma M).$$

So, $N = Q_\tau(M)/\overline{\sigma}Q_\tau(M) = Q_\tau(M)/Q_\tau(\sigma M)$ injects into $Q_\tau(M/\sigma M)$, and since $Q_\tau(M/\sigma M)$ is σ-torsionfree, so is N. Let $E(N)$ denote an injective hull of N in $Q_\tau(R)-\mathbf{mod}$, then $N \subseteq Q_{\overline{\sigma}}(N) \subseteq E(N)$. By construction $E(N)/Q_{\overline{\sigma}}(N)$ is $\overline{\sigma}$-torsionfree, hence σ-torsionfree and it easily follows that $Q_{\overline{\sigma}}(N)$ is σ-injective. Finally, the quotient $Q_{\overline{\sigma}}(N)/N$ is $\overline{\sigma}$-torsion, hence σ-torsion, so we obtain an isomorphism $Q_{\overline{\sigma}}(N) \cong Q_\sigma(N)$. To complete the proof, it suffices to note that $Q_{\overline{\sigma}}Q_\tau(M) = Q_{\overline{\sigma}}(N) \cong Q_\sigma(N) = Q_\sigma Q_\tau(M)$. $\square$

In order to prove the result announced in the introduction to this Section (i.e., the fact that σ is Q_τ-compatible if and only if σ and τ are mutually compatible), we will need the preliminary results below, which present some interest in their own right.

(3.4) Lemma. Let R be a ring and σ and τ two radicals in $R-\mathbf{mod}$. Then the following assertions are equivalent:

(3.4.1) $\mathcal{T}_\sigma$ is closed under Q_τ;

(3.4.2) If $M \in \mathcal{F}_\tau$, then $M/\sigma M \in \mathcal{F}_\tau$;

(3.4.3) $\mathcal{F}_\tau$ is closed under Q_σ.

Proof.   To prove that (1) implies (2), let $M \in \mathcal{F}_\tau$. By assumption, we have $Q_\tau(\sigma M) \in \mathcal{T}_\sigma$, so the canonical inclusion

$$\alpha : M/\sigma M \hookrightarrow Q_\tau(M)/\sigma Q_\tau(M)$$

fits into the commutative diagram

$$
\begin{array}{ccc}
M/\sigma M & \xrightarrow{\ \alpha\ } & Q_\tau(M)/\sigma Q_\tau(M) \\
\downarrow{\scriptstyle j_{M/\sigma M}} & & \uparrow \\
Q_\tau(M/\sigma M) & \longrightarrow & Q_\tau(M)/Q_\tau(\sigma M)
\end{array}
$$

Since α is injective, so is $j_{M/\sigma M}$, and this yields that $M/\sigma M$ is τ-torsionfree, indeed.

Conversely, we claim that (2) implies (1). For, if M is σ-torsion, then so is $M/\tau M$, and there is a surjection

$$
Q_\tau(M)/(M/\tau M) \to Q_\tau(M)/\sigma Q_\tau(M).
$$

Since $Q_\tau(M)/(M/\tau M)$ is τ-torsion, $Q_\tau(M)/\sigma Q_\tau(M)$ is τ-torsion as well. But $Q_\tau(M) \in \mathcal{F}_\tau$, so our assumptions imply $Q_\tau(M)/\sigma Q_\tau(M) \in \mathcal{F}_\tau$, and therefore $Q_\tau(M) = \sigma Q_\tau(M) \in \mathcal{T}_\sigma$.

Finally, the equivalence between (2) and (3) follows from from the fact that $\mathcal{F}_\tau$ is closed under taking submodules and essential extensions.

$\square$

It immediately follows:

(3.5) Corollary. *Let R be a ring and let σ and τ be radicals in $R-\mathbf{mod}$. Then the following assertions are equivalent:*

(3.5.1) $\mathcal{T}_\sigma$ *and $\mathcal{F}_\sigma$ are closed under Q_τ;*

(3.5.2) $\mathcal{T}_\tau$ *and $\mathcal{F}_\tau$ are closed under Q_σ.*

The above Lemma may be extended as follows:

(3.6) Lemma. *Let R be a ring and let σ and τ be radicals in $R-\mathbf{mod}$. Then the following assertions are equivalent:*

(3.6.1) $\mathcal{T}_\sigma$ *is closed under Q_τ;*

(3.6.2) $(R, \tau)-\mathbf{mod}$ *is closed under σ;*

(3.6.3) σ *induces (by restriction) a radical in $(R, \tau)-\mathbf{mod}$.*

Proof. To prove that (1) implies (2), consider a τ-closed left R-module M. Since $\sigma M \subseteq M$ is τ-torsionfree, and since $\mathcal{T}_\sigma$ is closed under Q_τ, we obtain inclusions

$$\sigma M \subseteq Q_\tau(\sigma M) \subseteq \sigma Q_\tau(M) = \sigma M,$$

proving that $Q_\tau(\sigma M) = \sigma M$. So, $(R, \tau)-\mathbf{mod}$ is closed under σ, indeed.

Conversely, assume (2) and let M be a σ-torsion left R-module. Since $\overline{M} = M/\tau M \in \mathcal{T}_\sigma$ as well, the canonical morphism

$$j_{\tau, \overline{M}} : \overline{M} \hookrightarrow Q_\tau(\overline{M}) = Q_\tau(M)$$

factorizes through $\sigma Q_\tau(M)$. As $\sigma Q_\tau(M)$ is τ-closed, by assumption, we thus obtain an (injective) extension $\overline{j_M} : Q_\tau(M) \hookrightarrow \sigma Q_\tau(M)$, so $Q_\tau(M) \in \mathcal{T}_\sigma$. This proves that (2) implies (1).

Finally, the equivalence between (2) and (3) is obvious. $\square$

The following result is the key to (3.8):

(3.7) Proposition. *Let R be a ring, and σ and τ two radicals in $R-\mathbf{mod}$. Then the following assertions are equivalent:*

(3.7.1) σ *is Q_τ-compatible;*

(3.7.2) $\mathcal{T}_\sigma$ *and $\mathcal{F}_\sigma$ are closed under Q_τ.*

Proof. That (1) implies (2) is obvious. Conversely, assume (2) and let M be a left R-module, then we have an exact commutative diagram

$$
\begin{array}{ccccccccc}
0 & \longrightarrow & \sigma M & \longrightarrow & M & \longrightarrow & M/\sigma M & \longrightarrow & 0 \\
& & \downarrow & & \downarrow & & \downarrow & & \\
0 & \longrightarrow & Q_\tau(\sigma M) & \longrightarrow & Q_\tau(M) & \longrightarrow & Q_\tau(M/\sigma M) & &
\end{array}
$$

By assumption, $Q_\tau(\sigma M) \subseteq \sigma Q_\tau(M)$. Moreover, $Q_\tau(M/\sigma M)$ is σ-torsionfree, hence so is

$$\sigma Q_\tau(M)/Q_\tau(\sigma M) \subseteq Q_\tau(M/\sigma M).$$

So, we have the other inclusion, as well, i.e., σ is Q_τ-compatible. $\quad\square$

Combining (3.5) and (3.7) now finally yields:

(3.8) Theorem. *Let R be a ring and σ and τ two radicals in $R-\mathbf{mod}$. Then the following assertions are equivalent:*

(3.8.1) *σ is Q_τ-compatible;*

(3.8.2) *τ is Q_σ-compatible;*

(3.8.3) *σ and τ are mutually compatible;*

(3.8.4) *$\mathcal{T}_\sigma$ and $\mathcal{F}_\sigma$ are closed under Q_τ;*

(3.8.5) *$(R,\tau)-\mathbf{mod}$ is closed under σ and $(R,\sigma)-\mathbf{mod}$ under τ;*

(3.8.6) *by restriction, σ induces a radical in $(R,\tau)-\mathbf{mod}$ and τ a radical in $(R,\sigma)-\mathbf{mod}$.*

(3.9) We will give in (3.11) still another interpretation of compatibility. In order to realize this, we will need some preliminaries first, however.

For any set A, we denote by $\mathcal{P}(A)$ the collection of its finite subsets. Recall from [71] that if $\mathcal{L}$ and $\mathcal{H}$ are two filters of left ideals of R, then the filter $\mathcal{L} \circ \mathcal{H}$ is defined as consisting of all left ideals L of R, with the property that there exists some $H \in \mathcal{H}$, containing L, and such that $(L : U) \in \mathcal{L}$, for any $U \in \mathcal{P}(H)$. It is easy to check that $(\mathcal{L} \circ \mathcal{H}) \circ \mathcal{K} = \mathcal{L} \circ (\mathcal{H} \circ \mathcal{K})$, when $\mathcal{L}$ is a cofilter, i.e., if $\mathcal{L}$ is also closed under taking intersections.

As in [71], let us call $\mathcal{L}$ *uniform*, if $\mathcal{L} \circ \{R\} \subseteq \mathcal{L}$. It is easy to verify that if $\mathcal{L}$ and $\mathcal{H}$ are uniform filters, then so is $\mathcal{L} \circ \mathcal{H}$. Moreover, if $\mathcal{L}$ is uniform, then it is unambiguously determined by $Cl_{\mathcal{L}}^R$, since, in this case, $I \in \mathcal{L}$ if and only if $Cl_{\mathcal{L}}^R(I) = R$. Here, for any pair of left R-modules $N \subseteq M$ and any filter of left R-ideals $\mathcal{L}$, we let $Cl_{\mathcal{L}}^M(N)$, the *$\mathcal{L}$-closure* of N in M, consist of all $m \in M$, such that $Lm \subseteq N$, for some $L \in \mathcal{L}$. We write this as $Cl_\sigma^M(N)$, if $\mathcal{L} = \mathcal{L}(\sigma)$ for some radical σ in $R-\mathbf{mod}$. Of course, if E is an injective hull of $M/\sigma M$, then clearly

$Q_\sigma(M) = Cl_\sigma^E(M/\sigma M)$.

With these definitions, it is fairly easy to check that a filter of left ideals $\mathcal{L}$ is a Gabriel filter if and only if it is uniform and $\mathcal{L} \circ \mathcal{L} \subseteq \mathcal{L}$. Note however, that if $\mathcal{L}(\sigma)$ and $\mathcal{L}(\tau)$ are Gabriel filters, then $\mathcal{L}(\sigma) \circ \mathcal{L}(\tau)$ is, of course, uniform, but not necessarily a Gabriel filter anymore. However, as we will see below, $\mathcal{L}(\sigma) \circ \mathcal{L}(\tau)$ is closely connected to $\mathcal{L}(\sigma \vee \tau)$, the Gabriel filter of the join $\sigma \vee \tau$ of σ and τ in the lattice $R-\mathbf{rad}$. Finally, we leave it as an easy exercise to the reader to verify that for any pair of radicals σ, τ in $R-\mathbf{mod}$, a left R-ideal I belongs to $\mathcal{L}(\sigma) \circ \mathcal{L}(\tau)$ if and only if there exists some $J \in \mathcal{L}(\tau)$ with $I \subseteq J$ and with $J/I \in \mathcal{T}_\sigma$.

As a first result, we may now prove:

(3.10) Lemma. *Let σ and τ be radicals in $R-\mathbf{mod}$. Then*

$$Cl_\sigma^M(Cl_\tau^M(N)) = Cl_{\mathcal{L}(\tau)\circ\mathcal{L}(\sigma)}^M(N)$$

for each pair of left R-modules $N \subseteq M$.

Proof. Let $x \in M$, then, clearly, $x \in Cl_\sigma^M(Cl_\tau^M(N))$ if and only if $I = (Cl_\tau^M(N) : x) \in \mathcal{L}(\sigma)$. So for all $i \in I$, we have $ix \in Cl_\tau^M(N)$, hence $(N : ix) \in \mathcal{L}(\tau)$. It thus follows that $((N : x) : j) \in \mathcal{L}(\tau)$ for each $j \in I + (N : x) \in \mathcal{L}(\sigma)$. Since $(N : x) \in \mathcal{L}(\tau) \circ \mathcal{L}(\sigma)$, by definition, it follows that $x \in Cl_{\mathcal{L}(\tau)\circ\mathcal{L}(\sigma)}^M(N)$, indeed.

Conversely, assume that $x \in Cl_{\mathcal{L}(\tau)\circ\mathcal{L}(\sigma)}^M(N)$, then there exists some $I \in \mathcal{L}(\tau) \circ \mathcal{L}(\sigma)$, such that $Ix \subseteq N$. But then there exists some $J \in \mathcal{L}(\sigma)$ with $I \subseteq J$ and such that $J/I \in \mathcal{T}_\tau$. This yields that $Jx \subseteq Cl_\tau^M(N)$, hence that $x \in Cl_\sigma^M(Cl_\tau^M(N))$. $\qquad\square$

(3.11) Proposition. *Let σ and τ be a pair of radicals in $R-\mathbf{mod}$. Then the following assertions are equivalent:*

(3.11.1) *σ, τ are mutually compatible;*

(3.11.2) *$Cl_\sigma^M Cl_\tau^M = Cl_\tau^M Cl_\sigma^M$, for any $M \in R-\mathbf{mod}$;*

(3.11.3) *$Cl_\sigma^R Cl_\tau^R = Cl_\tau^R Cl_\sigma^R$;*

(3.11.4) *if M is σ-torsionfree resp. τ-torsionfree then so is $M/\tau M$ resp. $M/\sigma M$;*

(3.11.5) $\mathcal{L}(\sigma) \circ \mathcal{L}(\tau)$ *is a Gabriel filter;*

(3.11.6) $\mathcal{L}(\sigma) \circ \mathcal{L}(\tau) = \mathcal{L}(\sigma \vee \tau)$;

(3.11.7) $\mathcal{L}(\sigma) \circ \mathcal{L}(\tau) = \mathcal{L}(\tau) \circ \mathcal{L}(\sigma)$;

(3.11.8) $(M/\sigma M)/\tau(M/\sigma M) = (M/\tau M)/\sigma(M/\tau M)$ *for all $M \in$ $R-$mod.*

Proof. To prove that (1) and (4) are equivalent, first assume (1) and take $M \in \mathcal{F}_\tau$. Applying the (left exact) functor Q_τ to the exact sequence

$$0 \to \sigma M \to M \to M/\sigma M \to 0,$$

and combining the injectivity of the morphisms $\alpha : \sigma M \to Q_\tau(\sigma M)$ and $\beta : M \to Q_\tau(M)$ with the Snake Lemma, it follows in a straightforward way that $Ker(\gamma : M/\sigma M \to Q_\tau(M/\sigma M))$ injects into $Coker(\alpha)$. By (1), it follows that $Q_\tau(\sigma M) \in \mathcal{T}_\sigma$, hence $Coker(\alpha) \in \mathcal{T}_\sigma$, so $Ker(\gamma) \in \mathcal{T}_\sigma$, as well. On the other hand, we also have that $Ker(\gamma) \in \mathcal{F}_\sigma$, so $Ker(\gamma) = 0$ and $M/\sigma M \in \mathcal{F}_\tau$.

On the other hand, if we assume that (4) holds, then $Q_\tau(M)/\sigma Q_\tau(M) \in \mathcal{F}_\sigma$, hence $Q_\tau(M)/\sigma Q_\tau(M) \in \mathcal{F}_{\sigma \vee \tau}$. So,

$$\sigma Q_\tau(M) = (\sigma \vee \tau)Q_\tau(M) = Q_\tau((\sigma \vee \tau)M) \supseteq Q_\tau(\sigma M).$$

On the other hand,

$$Q_\tau(M)/Q_\tau(\sigma M) \subseteq Q_\tau(M/\sigma M) \subseteq E((M/\sigma M)/\tau(M/\sigma M)) \in \mathcal{F}_\sigma$$

by (4). This yields that $\sigma Q_\tau(M) \subseteq Q_\tau(\sigma M)$, proving the equality $\sigma Q_\tau(M) = Q_\tau(\sigma M)$.

The equivalence of (2), (3) and (7) easily follows from the previous Lemma, bearing in mind that $Cl_\mathcal{L}^R = Cl_\mathcal{H}^R$ if and only if $\mathcal{L} = \mathcal{H}$, for any pair of uniform filters $\mathcal{L}$ and $\mathcal{H}$.

To prove that (3) and (7) are equivalent, it suffices to note that $L \in \mathcal{L}(\sigma) \circ \mathcal{L}(\tau)$ if and only if $I/L \in \mathcal{T}_\sigma$ for some $I \supseteq L$ belonging to $\mathcal{L}(\tau)$ or, equivalently, to $Cl_\sigma^R(L) \in \mathcal{L}(\tau)$, i.e., to $Cl_\tau^R(Cl_\sigma^R(L)) = R$.

To prove that (5) implies (6), assume $\mathcal{L}(\sigma) \circ \mathcal{L}(\tau)$ is a Gabriel filter and suppose that $\mathcal{L}(\sigma) \cup \mathcal{L}(\tau) \subseteq \mathcal{L}(\kappa)$ for some radical κ in $R-\mathbf{mod}$. If $I \in \mathcal{L}(\sigma) \circ \mathcal{L}(\tau)$, then there exists $J \in \mathcal{L}(\tau)$ such that $J/I \in \mathcal{T}_\sigma$. Hence, $J \in \mathcal{L}(\kappa)$ and $J/I \in \mathcal{T}_\kappa$, so $I \in \mathcal{L}(\kappa)$. Since $\mathcal{L}(\sigma) \circ \mathcal{L}(\tau)$ is a Gabriel filter, it thus follows that $\mathcal{L}(\sigma) \circ \mathcal{L}(\tau) = \mathcal{L}(\sigma \vee \tau)$.

As it is obvious that (6) implies (7), let us prove that (7) implies (5). By the above remarks, it suffices to verify that

$$(\mathcal{L}(\sigma) \circ \mathcal{L}(\tau)) \circ (\mathcal{L}(\sigma) \circ \mathcal{L}(\tau)) \subseteq \mathcal{L}(\sigma) \circ \mathcal{L}(\tau),$$

and this follows easily from (7) and the associativity properties of the operation $\circ$.

Next, assume (4) and let us derive (8). Let M be an arbitrary left R-module. For any $m \in M$, we have $\overline{m} \in Ker(M/\sigma M \to M/(\sigma \vee \tau)M)$ if and only if $Ann_R^\ell(m) \in \mathcal{L}(\sigma \vee \tau) = \mathcal{L}(\sigma) \circ \mathcal{L}(\tau)$. So there exists some $J \in \mathcal{L}(\tau)$ with $I \subseteq J$ and such that $J/I \in \mathcal{T}_\sigma$. In particular, $Jm \cong J/I \in \mathcal{T}_\sigma$, so $J\overline{m} = \overline{0}$, i.e., $\overline{m} \in \tau(M/\sigma M)$. From this it follows that

$$Ker(M/\sigma M \to M/(\sigma \vee \tau)M) = \tau(M/\sigma M),$$

hence that $(M/\sigma M)/\tau(M/\sigma M) = M/(\sigma \vee \tau)M$. Since, by symmetry, we also have $(M/\tau M)/\sigma(M/\tau M) = M/(\sigma \vee \tau)M$, this proves our assertion.

Finally, (8) implies (4), since, for example, for any $M \in \mathcal{F}_\sigma$, the assumption (8) implies that

$$M/\tau M = (M/\sigma M)/\tau(M/\sigma M) = (M/\tau M)/\sigma(M/\tau M) \in \mathcal{F}_\sigma.$$

This finishes the proof.   □

(3.12) If we work with symmetric radicals, then the previous notions may be given a more concrete interpretation.

Indeed, let us first define σ to be τ-*compatible*, if for all $I \in \mathcal{L}(\tau)$ and $J \in \mathcal{L}(\sigma)$, we may find $I' \in \mathcal{L}(\tau)$ and $J' \in \mathcal{L}(\sigma)$, with $I'J' \subseteq JI$.

As a first result, we then have the following property (which we mention without proof, as it will be considerably strengthened below):

(3.13) Proposition. [84] *Let R be left noetherian and let σ and τ be symmetric radicals in $R-$mod. Then the following assertions are equivalent:*

(3.13.1) *σ and τ are mutually compatible;*

(3.13.2) *σ is τ-compatible and τ is σ-compatible.*

(3.14) Let us assume R to be noetherian, for a moment. As in [90], define a radical σ in $R-$mod to be a *weak biradical*, if there exists a radical ρ in mod$-R$, with the property that $rad(\sigma(J/I)) = rad(\rho(J/I))$ for all twosided ideals $I \subseteq J$ of R. In the presence of the noetherian hypothesis, one easily shows that ρ may then be chosen to be symmetric, whenever σ is and that a symmetric radical σ in $R-$mod is a weak biradical if and only if for any twosided ideal K of R and any $I \in \mathcal{L}^2(\sigma)$, there exists some $J \in \mathcal{L}^2(\sigma)$ and a positive integer n with the property that $K^n J \subseteq IK$ and $JK^n \subseteq KI$, cf. [90, (5.30-31.)].

Using this, it is easy to prove:

(3.15) Proposition. *If σ is a symmetric radical in $R-$mod, then σ is a weak biradical if and only if if σ and σ^r are mutually compatible with any symmetric radical τ in $R-$mod resp. mod$-R$.*

Proof. The previous remarks clearly imply any weak biradical σ to possess the compatibility properties we claim, since for any $I \in \mathcal{L}^2(\sigma) = \mathcal{L}^2(\sigma^r)$ and any positive integer n, we have $I^n \in \mathcal{L}^2(\sigma)$, as well.

Conversely, assume σ and σ^r to be compatible with every symmetric radical τ (in the appropriate category of R-modules), then, in particular, for any twosided ideal K of R, they are compatible with the radical σ_K in $R-$**mod** and its right handed analogue $\rho_K = \sigma_K^r$ in **mod**$-R$. It follows that for any $I \in \mathcal{L}^2(\sigma)$, we may find $J \in \mathcal{L}^2(\sigma)$ and $H \in \mathcal{L}^2(\sigma_K) = \mathcal{L}^2(\rho_K)$, with $HJ \subseteq IK$ and $JH \subseteq KI$. But, by definition, H contains K^n for some positive integer n, hence we obtain $K^n J \subseteq IK$ and $JK^n \subseteq IK$. Applying the above remarks then yields the assertion. $\qquad\square$

The following results describe even more precisely how the different notions of compatibility are related:

(3.16) Proposition. *Let R be an arbitrary ring and let σ and τ be radicals in $R-$**mod**. If σ has finite type and if τ is symmetric, then σ is τ-compatible, whenever $\mathcal{T}_\sigma$ is closed under Q_τ.*

Proof. Consider some $I \in \mathcal{L}(\tau)$ and some $J \in \mathcal{L}(\sigma)$, together with the exact sequence

$$0 \to I/JI \to R/JI \to R/I \to 0.$$

Since $I \in \mathcal{L}(\tau)$, this yields an isomorphism $Q_\tau(I/JI) \cong Q_\tau(R/JI)$. Moreover, by assumption, as I/JI is σ-torsion, so is $Q_\tau(R/JI)$. Hence $(R/JI)/\tau(R/JI) \in \mathcal{T}_\sigma$, since it injects into $Q_\tau(R/JI)$. If we denote the image of $1 \in R$ in R/JI by $\overline{1}$, then it follows that $J'\overline{1} \in \tau(R/JI)$ for some $J' \in \mathcal{L}(\sigma)$, which we may, obviously, choose to be finitely generated. But then, since τ is symmetric, we may find some ideal $I' \in \mathcal{L}(\tau)$, with $I'J'\overline{1} = 0$, i.e., $I'J' \subseteq JI$. Indeed, it suffices to pick, for any j' in a finite set of generators of the left ideal J', some ideal $J_{j'} \in \mathcal{L}(\tau)$ annihilating $\overline{j'} = j'\overline{1}$, and then, to let I' be the intersection of these. This proves that $\mathcal{T}_\sigma$ is closed under Q_τ, indeed. $\qquad\square$

Conversely, we also have:

(3.17) Proposition. *Let R be an arbitrary ring and let σ and τ be radicals in $R-$**mod**. If τ has finite type and if σ is symmetric, then $\mathcal{T}_\sigma$*

is closed under Q_τ, whenever σ is τ-compatible.

Proof. Let M be a σ-torsion R-module and pick $m \in Q_\tau(M)$. We want to show that m belongs to $\sigma Q_\tau(M)$. Since the cokernel of the map $M/\tau M \to Q_\tau(M)$ is τ-torsion, we may find some finitely generated $I \in \mathcal{L}(\tau)$, with the property that $Im \subseteq M/\tau M$. Arguing as in the proof of the previous result, let us pick an ideal $J \in \mathcal{L}(\sigma)$, with $JIm = 0$. By assumption, there exist $I' \in \mathcal{L}(\tau)$ and $J' \in \mathcal{L}(\sigma)$, with $I'J' \subseteq JI$. So, $I'J'm = 0$, and $J'm \subseteq \tau Q_\tau(M) = 0$. Finally, the last statement yields that $m \in \sigma Q_\tau(M)$, indeed, which proves the assertion. $\square$

(3.18) Remark. In particular, the foregoing results show that if σ and τ are both symmetric radicals of finite type in $R-\mathbf{mod}$, then σ is τ-compatible, if and only if $\mathcal{T}_\sigma$ is closed under Q_τ. In the left noetherian case, this permits us to recover [84, Theorem (1.6.)], which is thus considerably strengthened by the previous results.

(3.19) As an easy application, let us briefly consider the behaviour of stable radicals in $R-\mathbf{mod}$ with respect to compatibility. Using the Artin-Rees property, it has been proved in [73], that if R is left noetherian, then any pair of stable symmetric radicals σ and τ in the category $R-\mathbf{mod}$ are mutually compatible. Moreover, using cohomological methods, it was deduced from this, in loc. cit., that under the same assumptions, Q_σ and Q_τ commute.

Let us show how these results may also be deduced (and even strengthened), just using the above results on compatibility.

(3.20) Lemma. *Let R be an arbitrary ring and let τ be a stable radical in $R-\mathbf{mod}$. Then, for any radical σ in $R-\mathbf{mod}$, the torsion class $\mathcal{T}_\tau$ is closed under Q_σ.*

Proof. If M belongs to $\mathcal{T}_\tau$, then $M/\sigma M$ is σ-torsionfree and τ-torsion, hence, so is $E(M/\sigma M)$. Moreover, since $E(M/\sigma M)$ is injective and σ-torsionfree, it is σ-closed, i.e., it belongs to $\mathcal{T}_\tau \cap (R, \sigma)-\mathbf{mod}$, within

$R-$**mod.** Localizing the inclusion $M/\sigma M \subseteq E(M/\sigma M)$ at σ yields an inclusion

$$Q_\sigma(M) = Q_\sigma(M/\sigma M) \subseteq Q_\sigma(E(M/\sigma M)) = E(M/\sigma M),$$

hence $Q_\sigma(M)$ is τ-torsion, as a submodule of $E(M/\sigma M) \in \mathcal{T}_\tau$. $\square$

(3.21) Corollary. *Any two stable radicals σ and τ in $R-$mod are mutually compatible.*

Proof. This follows easily from (3.8) and the foregoing Lemma.

$\square$

Let us now briefly investigate when localization functors commute. First note that it easily follows from (3.11), that for any pair of mutually compatible radicals in $R-$**mod**, the localization functors Q_σ and Q_τ commute on every $M \in \mathcal{T}_{\sigma \vee \tau}$.

For any left R-module M and any pair of radicals σ and τ in $R-$**mod**, we now define

$$C_\sigma^\tau(M) = Coker(Q_\tau(M) \to Q_\tau(M/\sigma M)).$$

We may then prove:

(3.22) Proposition. *Let M be a left R-module and let σ and τ radicals in $R-$mod. Then the following assertions are equivalent:*

(3.22.1) *for any left R-module M, we have $Q_\sigma Q_\tau(M) = Q_\tau Q_\sigma(M)$;*

(3.22.2) *σ and τ are mutually compatible and for any left R-module M, we have $C_\sigma^\tau(M) \in \mathcal{T}_\sigma$ and $C_\tau^\sigma(M) \in \mathcal{T}_\tau$.*

Proof. Let us first show that (1) implies (2). We claim that σ and τ are mutually compatible. Let $M \in \mathcal{T}_\sigma$, then $Q_\sigma(M) = 0$, and so $0 = Q_\tau Q_\sigma(M) = Q_\sigma Q_\tau(M)$. So, $Q_\tau(M) \in \mathcal{T}_\sigma$, hence, if we apply (3.8), we find that σ and τ are mutually compatible. Let us now show $C_\sigma^\tau(M)$ to be σ-torsion. For any left R-module M, we have an exact sequence

$$0 \to Q_\tau(\sigma M) \to Q_\tau(M) \to Q_\tau(M/\sigma M) \to C_\sigma^\tau(M) \to 0.$$

If we apply Q_σ to this sequence, we obtain the exact sequence

$$0 \to Q_\sigma Q_\tau(M) \xrightarrow{\alpha} Q_\sigma Q_\tau(M/\sigma M) \xrightarrow{p} Q_\sigma(C_\sigma^\tau(M)),$$

because $Q_\tau(\sigma M) = \sigma Q_\tau(M)$, as $Q_\sigma Q_\tau(M) = Q_\sigma(Q_\tau(M)/\sigma Q_\tau(M))$. Now since

$$Q_\sigma Q_\tau(M) = Q_\tau Q_\sigma(M) = Q_\tau Q_\sigma(M/\sigma M) = Q_\sigma Q_\tau(M/\sigma M),$$

α is an isomorphism, hence p is the zero-map. Since p fits into the commutative square

$$
\begin{array}{ccc}
Q_\tau(M/\sigma M) & \longrightarrow & C_\sigma^\tau(M) \\
\downarrow & & \downarrow{\scriptstyle j_{\sigma, C_\sigma^\tau(M)}} \\
Q_\sigma Q_\tau(M/\sigma M) & \xrightarrow{\ p\ } & Q_\sigma(C_\sigma^\tau(M))
\end{array}
$$

it follows that $j_{\sigma, C_\sigma^\tau(M)} : C_\sigma^\tau(M) \to Q_\sigma(C_\sigma^\tau(M))$ is the zero-map as well, showing that $C_\sigma^\tau(M)$ is σ-torsion, indeed.

Conversely, let us consider the exact sequences

$$0 \to \tau M \to M \to M/\tau M \to 0$$

resp.

$$0 \to M/\tau M \to Q_\tau(M) \to T_\tau(M) \to 0,$$

where $T_\tau(M) = Coker(j_{\tau,M})$ is τ-torsion. Applying the functor Q_σ to these two sequences, we obtain the exact sequences

$$0 \to Q_\sigma(\tau M) \to Q_\sigma(M) \xrightarrow{\beta} Q_\sigma(M/\tau M) \to C_\tau^\sigma(M) \to 0$$

resp.

$$0 \to Q_\sigma(M/\tau M) \xrightarrow{\gamma} Q_\sigma Q_\tau(M) \to Q_\sigma(T_\tau(M)).$$

Since τ is σ-stable, $Q_\sigma(\tau M)$ and $Q_\sigma(T_\tau(M))$ are τ-torsion. On the other hand, by assumption, $C_\tau^\sigma(M)$ is τ-torsion, so β and γ are τ-isomorphisms, hence so is $\gamma\beta : Q_\sigma(M) \to Q_\sigma Q_\tau(M)$. We thus find that

$Q_\tau Q_\sigma(M) = Q_\tau Q_\sigma Q_\tau(M)$ (up to isomorphism). Now, since $Q_\tau(M) \in \mathcal{F}_\tau$, the canonical homomorphism

$$j_{\tau, Q_\sigma Q_\tau(M)} : Q_\sigma Q_\tau(M) \to Q_\tau Q_\sigma Q_\tau(M)$$

is injective, so we find an injection $Q_\sigma Q_\tau(M) \hookrightarrow Q_\tau Q_\sigma(M)$. In a similar way, one constructs an injection $Q_\tau Q_\sigma(M) \hookrightarrow Q_\sigma Q_\tau(M)$. Since both of these extend the identity on M, a straightforward unicity argument easily shows them to be inverse to each other, which finishes the proof.

$\square$

(3.23) Corollary. *If σ and τ are stable radicals in $R-$**mod**, then* $Q_\sigma Q_\tau = Q_\tau Q_\sigma$.

Proof. Since σ and τ are stable, it follows from (3.21) that σ and τ are mutually compatible.

On the other hand, we claim that $C_\sigma^\tau(M) \in \mathcal{T}_\sigma$. Indeed, obviously $C_\sigma^\tau(M)$ injects into $R^1 Q_\tau(\sigma M)$. Now, σM is σ-torsion and as σ is stable, σM possesses an injective resolution $E_\bullet$, all of whose terms are σ-torsion. Since every $Q_\tau(E_i)$ is σ-torsion (as σ and τ are mutually compatible), obviously so is $R^1 Q_\tau(\sigma M)$, proving that $C_\sigma^\tau(M) \in \mathcal{T}_\sigma$, indeed. By symmetry, $C_\tau^\sigma(M) \in \mathcal{T}_\tau$. So, σ and τ satisfy the conditions in the previous Proposition, which proves the assertion. $\square$

Note also, in the same context, that we have:

(3.24) Corollary. *If σ and τ are stable symmetric radicals in $R-$**mod**, then* $Q_\sigma Q_\tau = Q_\tau Q_\sigma = Q_{\sigma \vee \tau}$.

Proof. From the previous result, it easily follows that Q_σ and Q_τ commute and that $Q_\sigma Q_\tau$ is the localization functor, with respect to some radical ρ in $R-$**mod**. To prove the assertion, it remains to show that $\rho = \sigma \vee \tau$, or, equivalently, that $\mathcal{T}_{\sigma \vee \tau} = Ker(Q_\tau Q_\sigma)$, the subcategory of $R-$**mod**, consisting of all $M \in R-$**mod** with $Q_\tau Q_\sigma(M) = 0$. Now, $Q_\tau Q_\sigma(M) = 0$ is equivalent to $Q_\sigma(M) \in \mathcal{T}_\tau$, hence to $M/\sigma M \in \mathcal{T}_\tau$, since τ is stable. But, this just says that for any $m \in M$, there exists some $J \in \mathcal{L}(\tau)$ with $Jm \subseteq \sigma M$. It follows that $IJm = 0$, for some $I \in \mathcal{L}(\sigma)$, since J is finitely generated, R being left noetherian, and since τ is symmetric. Hence, $m \in (\sigma \vee \tau)M$, by (3.13) in the previous Chapter, so $M \in \mathcal{T}_{\sigma\vee\tau}$, indeed. $\square$

If R is a ring, M a left R-module and if σ and τ are radicals in $R-\mathbf{mod}$, then we say that σ and τ are *mutually compatible on* M, if $Q_\sigma(\tau M) = \tau Q_\sigma(M)$ and $Q_\tau(\sigma M) = \sigma Q_\tau(M)$. In particular, σ and τ will be mutually compatible if they are mutually compatible on every left R-module M.

As an example, let us show that if R is a left noetherian ring, then any pair of radicals in $R-\mathbf{mod}$, satisfying the Artin-Rees property are mutually compatible on normalizing R-bimodules.

We will need the following straightforward result:

(3.25) Lemma. *Let R be a left noetherian ring and let σ and τ be radicals in $R-\mathbf{mod}$. If σ satisfies the Artin-Rees property and if M is a normalizing R-bimodule, then $Q_\tau(\sigma M) \in \mathcal{T}_\sigma$.*

Proof. This follows from the observation that

$$Q_\tau(\sigma M) = Q_\tau(\sigma M/\tau(\sigma M)) \subseteq Q_\tau(\sigma(M/\tau M)) \subseteq E(\sigma(M/\tau M)),$$

where $E(\sigma(M/\tau M)) \in \mathcal{T}_\sigma$, as $E(\sigma(M/\tau M)) = \sigma E(M/\tau M)$, by (2.20) in the previous Chapter. $\square$

(3.26) Corollary. *Let R be a left noetherian ring and let σ and τ be radicals in $R-\mathbf{mod}$, which both satisfy the Artin-Rees property. Then σ and τ are mutually compatible on every normalizing R-bimodule.*

Proof. It suffices to verify that $Q_\sigma(\tau M) = \tau Q_\sigma(M)$, for any normalizing R-bimodule M. First, assume M to be σ-torsionfree, then we claim that $M/\tau M$ is σ-torsionfree, as well. Consider the following exact, commutative diagram:

$$
\begin{array}{ccccccccc}
0 & \longrightarrow & \tau M & \longrightarrow & M & \stackrel{p}{\longrightarrow} & M/\tau M & \longrightarrow & 0 \\
 & & \downarrow{\scriptstyle i} & & \downarrow{\scriptstyle j} & & \downarrow{\scriptstyle k} & & \\
0 & \longrightarrow & Q_\sigma(\tau M) & \longrightarrow & Q_\sigma(M) & \underset{q}{\longrightarrow} & Q_\sigma(M/\tau M) & &
\end{array}
$$

Since $M \in \mathcal{F}_\sigma$, clearly i and j are injective. To prove our assertion, we want to show k to be injective. So, let $m \in M$ and suppose that $k(p(n)) = 0$, then $j(m) \in Q_\sigma(\tau M) = Ker(q)$. Since $Q_\sigma(\tau M) \in \mathcal{T}_\tau$, by the previous Lemma, there exists some $L \in \mathcal{L}(\tau)$, with the property that $j(Lm) = Lj(m) = 0$. But then $Lm = 0$, as j is injective, so $m \in \tau N$, i.e., $p(m) = 0$, indeed.

Next, still assuming M to be normalizing and σ-torsionfree, it is easy to see that $Q_\tau(M) \in \mathcal{F}_\sigma$. Indeed, by the previous remarks, $M/\tau M \in \mathcal{F}_\sigma$, so $Q_\tau(M) \subseteq E(M/\tau M) \in \mathcal{F}_\sigma$.

Finally, assume M to be an arbitrary normalizing R-bimodule and consider the following exact sequence:

$$ 0 \to \sigma M \to M \to M/\sigma M \to 0 $$

which induces

$$ 0 \to \sigma Q_\tau(\sigma M) \to \sigma Q_\tau(M) \to \sigma Q_\tau(M/\sigma M). $$

Now, $\sigma Q_\tau(M/\sigma M) = 0$, by the preceding remarks, while $\sigma Q_\tau(\sigma M) = Q_\tau(\sigma M)$, by the foregoing Lemma. So, $Q_\tau(\sigma M) = \sigma Q_\tau(M)$, indeed.

$\square$

With this terminology, as we will see below, the need for compatibility in the context of sheaf constructions mainly stems from the next result:

(3.27) Theorem. *Let R be a ring, let M be a left R-module and let σ, τ be radicals in $R-\mathbf{mod}$. The following assertions are equivalent:*

(3.27.1) *the sequence*

$$0 \to Q_{\sigma \wedge \tau}(M) \to Q_\sigma(M) \oplus Q_\tau(M) \to Q_{\sigma \vee \tau}(M) \qquad \text{(i)}$$

is exact;

(3.27.2) $(\sigma \vee \tau)Q_\sigma Q_\tau(M) = 0$ and $(\sigma \vee \tau)Q_\tau Q_\sigma(M) = 0$;

(3.27.3) σ and τ are mutually compatible on M.

Proof. The homomorphisms in (i) are, of course, induced by localization. Let us verify that the first statement is equivalent to the exactness of the sequences

$$0 \to Q_\sigma(M) \to Q_\sigma(M) \oplus Q_\sigma Q_\tau(M) \to Q_{\sigma \vee \tau}(M) \qquad \text{(ii)}$$

resp.

$$0 \to Q_\tau(M) \to Q_\tau Q_\sigma(M) \oplus Q_\tau(M) \to Q_{\sigma \vee \tau}(M), \qquad \text{(iii)}$$

where the homomorphisms are still the natural ones.

Of course, (ii) and (iii) are exact provided (i) is, since Q_σ and Q_τ are left exact. On the other hand, assume (ii) to be exact and consider the exact sequence

$$0 \to K \to Q_\sigma(M) \oplus Q_\tau(M) \to Q_{\sigma \vee \tau}(M). \qquad \text{(iv)}$$

If we apply $Q_{\sigma \wedge \tau}$ to (iv), we easily obtain that K is $(\sigma \wedge \tau)$-closed. Let $j : Q_{\sigma \wedge \tau}(M) \hookrightarrow K$ denote the canonical injection. Applying Q_σ resp. Q_τ, to (iv) and using the exactness of (ii), we then deduce that $Q_\sigma(j)$ and $Q_\tau(j)$ are isomorphisms. So, $Coker(j) \in \mathcal{T}_{\sigma \wedge \tau}$ and as both $Q_{\sigma \wedge \tau}(M)$ and K are $(\sigma \wedge \tau)$-closed, it follows that j is an isomorphism, so (i) is exact.

Let us first prove that (1) implies (2), so, let us assume (ii) to be exact. The homomorphisms involved in (ii) are:

$$\varphi : Q_\sigma(M) \to Q_\sigma(M) \oplus Q_\sigma Q_\tau(M) : x \mapsto (x, \alpha(x)),$$

resp.

$$\psi : Q_\sigma(M) \oplus Q_\sigma Q_\tau(M) \to Q_{\sigma \vee \tau}(M) : (y, z) \mapsto \beta(y) - \gamma(z),$$

where α, β and γ are the obvious maps. Applying $Q_\sigma Q_\tau$ to the exact sequence

$$0 \to (\sigma \vee \tau)M \to M \to Q_{\sigma \vee \tau}(M),$$

we obtain that $(\sigma \vee \tau)Q_\sigma Q_\tau(M) = Q_\sigma Q_\tau((\sigma \vee \tau)M) = ker(\gamma)$. If we choose some $z \in (\sigma \vee \tau)Q_\sigma Q_\tau(M) = ker(\gamma)$, then $(0, z) \in ker(\psi)$ and $(0, z) = \varphi(x)$, for some $x \in Q_\sigma(M)$. It follows that $z = 0$, hence $(\sigma \vee \tau)Q_\sigma Q_\tau(M) = 0$ and, by symmetry (using the exactness of (iii)) that $(\sigma \vee \tau)Q_\tau Q_\sigma(M) = 0$.

Next let us prove that (2) implies (3). Since $(\sigma \vee \tau)Q_\tau Q_\sigma(M) = 0$, clearly $Q_\tau Q_\sigma(M) \in \mathcal{F}_{\sigma \vee \tau}$, i.e., $Q_\sigma(M)/\tau Q_\sigma(M) \in \mathcal{F}_{\sigma \vee \tau}$. So,

$$Q_\sigma(\tau M) \subseteq Q_\sigma((\sigma \vee \tau)M) = (\sigma \vee \tau)Q_\sigma(M) = \tau Q_\sigma(M).$$

On the other hand, our assumptions are easily seen to imply that $\mathcal{F}_\tau$ is closed under Q_σ, so from $Q_\sigma(M)/Q_\sigma(\tau M) \hookrightarrow Q_\sigma(M/\tau M)$, it follows that $Q_\sigma(M)/Q_\sigma(\tau M) \in \mathcal{F}_\sigma$, therefore $\tau Q_\sigma(M) \subseteq Q_\sigma(\tau M)$.

Finally, assume (3), i.e., suppose that $\tau Q_\sigma(M) = Q_\sigma(\tau M)$, and let us prove (1). It is clear that $(\sigma \vee \tau)Q_\tau Q_\sigma(M) = 0$. Now, if we take $(y, z) \in ker(\psi)$, then $0 = \beta(y) - \gamma(z) = \gamma(\alpha(y) - z)$, so $\alpha(y) - z \in ker(\gamma) = 0$. Hence $(y, z) = \varphi(y)$ and (ii) is exact. A symmetric argument also shows (iii) to be exact, which finishes the proof. $\qquad\square$

4. Topologies

(4.1) As before, let us associate to any radical σ in $R-$**mod** the set $\mathcal{K}(\sigma)$ consisting of all $P \in Spec(R)$, with $R/P \in \mathcal{F}_\sigma$, cf. (4.8) in Chapter I. We wish to use these sets as open subsets for topologies on $Spec(R)$. So, let us first study in some detail topologies induced on $Spec(R)$ by families of radicals in $R-$**mod**. Let us call a radical σ in $R-$**mod** *disjunctive*, if it has the property that for any $P \in Spec(R)$, we have $P \in \mathcal{K}(\sigma)$ if and only if $P \notin \mathcal{L}(\sigma)$.

(4.2) Lemma. *Let G be a family of radicals in $R-$**mod**. Then*

$$\mathcal{K}(\bigvee G) = \bigcap_{\sigma \in G} \mathcal{K}(\sigma).$$

If every $\sigma \in G$ is disjunctive, then we also have that

$$\mathcal{K}(\bigwedge G) = \bigcup_{\sigma \in G} \mathcal{K}(\sigma).$$

Proof. The first statement follows immediately from the fact that

$$\mathcal{F}_{\vee G} = \bigcap_{\sigma \in G} \mathcal{F}_\sigma.$$

We also have that $\bigcup_{\sigma \in G} \mathcal{K}(\sigma) \subseteq \mathcal{K}(\bigwedge G)$, as $\sigma(R/P) = 0$ for some $\sigma \in G$ implies that $(\bigwedge G)(R/P) = \bigcap_{\sigma \in G} \sigma(R/P) = 0$. Finally, if every $\sigma \in G$ is disjunctive and if $P \in \mathcal{K}(\bigwedge G)$, then

$$R/P \notin \mathcal{T}_{\wedge G} = \bigcap_{\sigma \in G} \mathcal{T}_\sigma,$$

i.e., R/P is not σ-torsion for some $\sigma \in G$ or, equivalenty, $P \in \mathcal{K}(\sigma)$. This proves the assertion.   □

The previous property applies in particular, when all $\sigma \in G$ are symmetric, or when for every $P \in \mathcal{K}(\sigma)$ the quotient R/P is a left Goldie ring (e.g., if R is left noetherian), cf. (4.9) in Chapter I.

(4.3) From the foregoing Lemma, it follows that for any set $\mathcal{S}$ of disjunctive radicals in $R-$**mod**, closed under taking arbitrary meets and finite joins, the set

$$T(\mathcal{S}) = \{\mathcal{K}(\sigma);\ \sigma \in \mathcal{S}\}$$

is a topology on $Spec(R)$.

The most typical example of this situation is, of course, given by considering a left noetherian ring R and the set $\mathcal{S}_{zar}(R)$ of all (symmetric) radicals in $R-$**mod** of the form σ_I, for some twosided ideal I of R. In this case, $\mathcal{K}(\sigma_I) = X_R(I)$, the set of all prime ideals P of R with $I \not\subseteq P$. So, the associated topology $T_{zar}(R) = T(\mathcal{S}_{zar}(R))$ is just the traditional Zariski topology on $Spec(R)$.

Other useful choices for $\mathcal{S}$ have been studied in the literature. Let us give a brief survey here and refer to the references at the end for more details.

(4.4) Example. Let R be left noetherian. We have seen in (1.6) in the previous Chapter, that the family $\mathcal{S}_s(R)$ of all *stable* radicals in $R-$**mod** is closed under taking finite joins and arbitrary meets (and *arbitrary* joins if R is left noetherian). So, this defines the *stable* topology $T_s(R) = T(\mathcal{S}_s(R))$ on $Spec(R)$.

In a similar way, if R is noetherian, then the family $\mathcal{S}_s^2(R)$ of *bistable* radicals in $R-$**mod** (i.e., with the property that both σ and σ^r are stable) yields a topology, the so-called *bistable* topology $T_s^2(R) = T(\mathcal{S}_s^2(R))$ on $Spec(R)$.

(4.5) Example. We have seen in (3.12) and (3.14) in the previous Chapter, that the set $\mathcal{S}_w(R)$ of all symmetric radicals in $R-$**mod** satisfying the weak Artin-Rees property is closed under taking finite joins and arbitrary meets. So, this defines the so-called *weak Artin-Rees* topology $T_w(R) = T(\mathcal{S}_w(R))$ on $Spec(R)$.

Similarly, if R is noetherian, then the family $\mathcal{S}_w^2(R)$ consisting of all symmetric radicals σ in $R-\mathbf{mod}$ with the property that both σ and σ^r satisfy the weak Artin-Rees property in their respective categories of modules yields a topology $T_w^2(R) = T(\mathcal{S}_w^2(R))$ on $Spec(R)$.

(4.6) Example. As we pointed out in (4.5) in the previous Chapter, the set $\mathcal{S}_t(R)$ of all biradicals in $R-\mathbf{mod}$ is a complete distributive lattice. In particular, $\mathcal{S}_t(R)$ is closed under taking arbitrary meets and joins and hence yields a topology $T_t(R) = T(\mathcal{S}_t(R))$ on $Spec(R)$, the so-called *biradical* topology.

(4.7) Example. The same proof as the one given in (4.5) in Chapter III, shows that the sets $\mathcal{S}_{tc}(R)$ resp. $\mathcal{S}_{tn}(R)$ of all centralizing resp. strongly normalizing biradicals in $R-\mathbf{mod}$ form a complete distributive lattice, hence yield topologies $T_{tc}(R) = T(\mathcal{S}_{tc}(R))$ resp. $T_{tn}(R) = T(\mathcal{S}_{tn}(R))$ on $Spec(R)$.

(4.8) In general, if we assume R to be noetherian, for simplicity's sake, it appears that $T_{tc}(R)$ and its Zariski counterpart $T(R) = T_{tc}(R) \cap T_{zar}(R)$ will be the most useful candidates for a topology on $Spec(R)$.

Indeed, whereas most of the topologies introduced above yield structure *sheaves* on $Spec(R)$, as we are about to see, these structure sheaves fail to behave functorially, in general – except when constructed over $T_{tc}(R)$ or $T(R)$.

On the other hand, we should point out that, due to (4.13) in Chapter III, the topologies $T_t(R)$, $T_{tc}(R)$ and $T_{tn}(R)$ coincide, whenever R satisfies the strong second layer condition. Under the same assumption, by (4.10) in Chapter III, so do the intersections of $T_s^2(R)$, $T_w^2(R)$, $T_t(R)$ (and hence $T_{tc}(R)$, $T_{tn}(R)$) with the Zariski topology.

(4.9) Let us now fix a set $\mathcal{S}$ of radicals in $R-\mathbf{mod}$. If R is left noetherian and if $\mathcal{S}$ consists of symmetric radicals, then we know that for any $\sigma \in \mathcal{S}$, the set $\mathcal{K}(\sigma)$ and σ determine each other unambiguously. In particular, it then follows for any pair of radicals $\sigma, \tau \in \mathcal{S}$, that $\sigma \leq \tau$ if and only if $\mathcal{K}(\sigma) \supseteq \mathcal{K}(\tau)$.

In this case, it is clear that for any left R-module M, associating $Q_\sigma(M)$ to $\mathcal{K}(\sigma)$ for any $\sigma \in \mathcal{S}$ defines a presheaf $\mathcal{O}_M$ on $(Spec(R), T(\mathcal{S}))$, which is easily seen to be separated.

(4.10) Somewhat more generally, assume we are given an order inverting lattice morphism $\gamma : T(\mathcal{S}) \to R-\mathbf{rad}$, i.e., a map which to any $K = \mathcal{K}(\sigma) \in T(\sigma)$ associates a radical $\gamma(K)$ in $R-\mathbf{rad}$, with the property that $K \subseteq K'$ implies $\gamma(K) \geq \gamma(K')$. Then one may still define for any $M \in R-\mathbf{mod}$ a presheaf $\mathcal{O}_M^\gamma$ on $(Spec(R), T(\mathcal{S}))$, by mapping $K = \mathcal{K}(\sigma) \in T(\mathcal{S})$ to $Q_{\gamma(K)}(M)$, for any $\sigma \in \mathcal{S}$.

Let us say that γ is *continuous* if, moreover, for any family $\{K_\alpha;\ \alpha \in A\}$ in $T(\mathcal{S})$, we have

$$\bigwedge_{\alpha \in A} \gamma(K_\alpha) = \gamma(\bigcup_{\alpha \in A} K_\alpha).$$

We then obtain:

(4.11) Proposition. *If* $\gamma : T(\mathcal{S}) \to R-\mathbf{rad}$ *is a continuous order inverting lattice morphism, then the presheaf* $\mathcal{O}_M^\gamma$ *on* $(Spec(R), T(\mathcal{S}))$ *is separated for any left R-module M.*

Proof. Assume that

$$\mathcal{K}(\sigma) = \bigcup_{\alpha \in A} \mathcal{K}(\sigma_\alpha) = \mathcal{K}(\bigwedge_{\alpha \in A} \sigma_\alpha),$$

for some family of radicals $\{\sigma_\alpha;\ \alpha \in A\}$ in $\mathcal{S}$, then we have to prove the map

$$res : Q_{\gamma(\mathcal{K}(\sigma))}(M) \to \prod_{\alpha \in A} Q_{\gamma(\mathcal{K}(\sigma_\alpha))}(M)$$

to be injective. Now, it is easy to see that for any $\alpha \in A$, there is an exact sequence of left R-modules

$$0 \to \gamma(\mathcal{K}(\sigma_\alpha))Q_{\gamma(\mathcal{K}(\sigma))}(M) \to Q_{\gamma(\mathcal{K}(\sigma))}(M) \overset{res_\alpha}{\to} Q_{\gamma(\mathcal{K}(\sigma_\alpha))}(M).$$

So, we obtain that $m \in Ker(res)$ if and only if

$$m \in \bigcap_{\alpha \in A} Ker(res_\alpha) = \bigcap_{\alpha \in A} \gamma(\mathcal{K}(\sigma_\alpha))Q_{\gamma(\mathcal{K}(\sigma))}(M)$$

$$= (\bigwedge_{\alpha \in A} \gamma(\mathcal{K}(\sigma_\alpha)))Q_{\gamma(\mathcal{K}(\sigma))}(M)$$
$$= \gamma(\bigcup_{\alpha \in A} \mathcal{K}(\sigma_\alpha))Q_{\gamma(\mathcal{K}(\sigma))}(M)$$
$$= \gamma(\mathcal{K}(\sigma))Q_{\gamma(\mathcal{K}(\sigma))}(M) = 0,$$

so the map *res* is injective, indeed.   □

(4.12) To see when $\mathcal{O}_M$ or, more generally, $\mathcal{O}_M^\gamma$ is a sheaf, recall that two radicals σ and τ in $R-\mathbf{mod}$ are said to be mutually compatible on some left R-module M, if $Q_\sigma(\tau M) = \tau Q_\sigma(M)$ and $Q_\tau(\sigma M) = \sigma Q_\tau(M)$. It has been proved in (3.27), that this is equivalent to asserting that the sequence

$$0 \to Q_{\sigma \wedge \tau}(M) \to Q_\sigma(M) \oplus Q_\tau(M) \to Q_{\sigma \vee \tau}(M)$$

be exact.

It thus follows that if R is left noetherian, then for any left R-module and any family $\mathcal{S}$ of symmetric radicals in $R-\mathbf{mod}$, which is closed under taking arbitrary meets and finite joins and such that every open set $\mathcal{K}(\sigma)$ is quasicompact in the topology $T(\mathcal{S})$, the presheaf $\mathcal{O}_M$ on $(Spec(R), T(\mathcal{S}))$ is a sheaf if and only if $\mathcal{S}$ consists of radicals, which are pairwise mutually compatible on M. We will then say that $\mathcal{S}$ resp. $T(\mathcal{S})$ is *compatible* with M. If $\mathcal{S}$ is compatible with *every* $M \in R-\mathbf{mod}$, i.e., if any two radicals in $\mathcal{S}$ are mutually compatible, then we will say that $\mathcal{S}$ resp. $T(\mathcal{S})$ is *compatible*.

It is easy to verify, still assuming R to be left noetherian, that there exists a largest topology on $Spec(R)$ resp. a largest subtopology of the Zariski topology on $Spec(R)$ which is compatible. Indeed, this is a straightforward consequence of Zorn's Lemma and the result below:

(4.13) Lemma. *If the radicals σ_1, σ_2 and τ in $R-\mathbf{mod}$ are mutually compatible, then so are $\sigma_1 \vee \sigma_2$ resp. $\sigma_1 \wedge \sigma_2$ and τ.*

Proof. This follows easily from the characterizations (3.11) of mutual compatibility, taking into account the fact that for any $M \in R-\mathbf{mod}$, we always have

$$(\sigma_1 \wedge \sigma_2)M = \sigma_1 M \cap \sigma_2 M$$

resp.

$$M/(\sigma_1 \vee \sigma_2)M = (M/\sigma_1 M)/\sigma_2(M/\sigma_1 M),$$

if σ_1 and σ_2 are mutually compatible. $\square$

Somewhat more generally, we also have the following:

(4.14) Proposition. *Let M be a left R-module, let $\mathcal{S}$ be a family of disjunctive radicals in $R-\mathbf{mod}$, which is closed under taking arbitrary meets and finite joins and with the property that every open subset $\mathcal{K}(\sigma)$ is quasicompact in the topology $T(\mathcal{S})$. Let $\gamma : T(\mathcal{S}) \to R-\mathbf{rad}$ be a continuous, order inverting lattice morphism. Then the presheaf $\mathcal{O}_M^\gamma$ on $(Spec(R), T(\mathcal{S}))$ is a sheaf if and only if for any pair of subsets $K, K' \in T(\mathcal{S})$, the radicals $\gamma(K)$ and $\gamma(K')$ are mutually compatible on M.*

5. Structure sheaves

(5.1) As we have already indicated in the introduction to this Chapter, our principal aim is to endow the prime spectrum of a reasonably general ring R with a structure sheaf, which behaves functorially with respect to central extensions, say. Several constructions of this type have been given in the past, the most successful one, of course, occurring in the commutative case, leading to the classical structure sheaves on the prime spectrum of a ring.

Both in [83] and [87] constructions have been given (in the noetherian case), which behaved well in some respects and badly in others. For example, the structure sheaf on $Spec(R)$ described in [83] is constructed very elegantly by means of localization at symmetric radicals, but unfortunately, does not behave functorially and actually only works in the prime case. (Of course, the latter restriction is not very serious, as it really just amounts to restrict to irreducible affine schemes). On the other hand, the constructions in [87] make use of some tricky type of localization within the (non-abelian!) category of centralizing R-bimodules and essentially only works nicely within the framework of pi rings, where they do, indeed, yield functorial structure sheaves.

In this Section, using the techniques developed in the previous Chapters, we will introduce structure sheaves on $Spec(R)$ (endowed with the topology $T(R)$), which represent the base ring R, behave functorially and which function nicely in a much wider class of rings.

(5.2) *In this Section, we will assume throughout R to be a noetherian ring.* As pointed out before, we may endow $Spec(R)$ with the subtopology $T(R)$ of the Zariski topology, consisting of all open subsets $X_R(I)$ with the property that σ_I is a centralizing biradical in $R-\mathbf{mod}$.

From results in the previous Section, it follows that associating $Q_I(R)$ to the open subset $X_R(I) \in T(R)$ defines a presheaf of rings $\mathcal{O}_R$ on $(Spec(R), T(R))$. Somewhat more generally, if we start from an arbitrary left R-module M, then the same construction, i.e., associating $Q_I(M)$ to $X_R(I) \in T(R)$, yields a sheaf of left $\mathcal{O}_R$-modules $\mathcal{O}_M$ over the ringed space $(Spec(R), T(R), \mathcal{O}_R)$. Of course, it follows directly from the definitions that $\mathcal{O}_M(Spec(R)) = M$, i.e., M may be recovered from $\mathcal{O}_M$ as the module of global sections. In particular, $\mathcal{O}_R(Spec(R)) = R$, so the structure sheaf $\mathcal{O}_R$ certainly satisfies our primary requirement of representing the ring R.

(5.3) A subset Y of a topological space X is said to be *closed under generization* or *generically closed*, if for any $x \in X$, we have that $x \in Y$ if and only if its closure $\overline{\{x\}}$ intersects Y non-trivially. For example, any open subset of X is obviously generically closed. In fact, one easily verifies that generically closed subsets are actually just arbitrary intersections of open subsets.

Obviously, as pointed out before, a subset Y of $Spec(R)$ endowed with the Zariski topology is closed under generization, if for any prime ideal Q of R, we have $Q \in Y$, whenever $Q \subseteq P$ for some $P \in Y$. If $Spec(R)$ is endowed with the topology $T(R)$, for example, then it is clear that the generically closed subsets of $(Spec(R), T(R))$ are just the subsets of the form $Y = \mathcal{K}(\sigma)$, where $\sigma = \bigvee \{\sigma_I \in \mathcal{S}_{tc}(R); \ Y \subseteq X_R(I) = \mathcal{K}(\sigma_I)\}$.

Taking sections for one of the sheaves defined above over an arbitrary generically closed subset of $(Spec(R), T(R))$ may be given the following torsion-theoretic interpretation:

(5.4) Proposition. *Let σ be a symmetric in $\mathcal{S}_{tc}(R)$, i.e., let $Y = \mathcal{K}(\sigma)$ be a generically closed subset of $(Spec(R), T(R))$. If M is a left R-module, then $(\mathcal{O}_M|Y)(Y) = Q_\sigma(M)$.*

Proof. For any twosided ideal I of R, we put $\tau_I = \sigma \vee \sigma_I$. Let us define a presheaf $\mathcal{F}$ of left R-modules on Y (endowed with the induced topology), by letting $\mathcal{F}(Y_R(I)) = Q_{\tau_I}(M)$, for any

$$Y_R(I) = X_R(I) \cap \mathcal{K}(\sigma) = \mathcal{K}(\tau_I).$$

The restriction maps for $\mathcal{F}$ are the canonical morphisms $Q_{\tau_I}(M) \to Q_{\tau_J}(M)$ deduced from $\tau_I \leq \tau_J$, whenever $Y_R(I) \supseteq Y_R(J)$.

Let us first note that $\mathcal{F}$ is a *separated* presheaf, i.e., that the map

$$\mathcal{F}(Y_R(\sum_{\alpha \in A} I_\alpha)) = \mathcal{F}(\bigcup_{\alpha \in A} Y_R(I_\alpha)) \to \prod_{\alpha \in A} \mathcal{F}(Y_R(I_\alpha))$$

is injective, for any family of twosided ideals $\{I_\alpha;\ \alpha \in A\}$ of R. Indeed, by the definition of $\mathcal{F}$, the kernel of this map is exactly

$$(\bigwedge_{\alpha \in A} \tau_{I_\alpha}) Q_{\sum_{\alpha \in A} I_\alpha}(M) = 0,$$

since it is easy to see that $\bigwedge_{\alpha \in A} \tau_{I_\alpha} = \tau_{\sum_{\alpha \in A} I_\alpha}$. So, $\mathcal{F}$ is separated, indeed.

Next, we claim that $\mathcal{F}$ is a sheaf. Using the noetherian assumption, one easily reduces the proof to verifying that for any pair of open subsets $Y_R(I)$ and $Y_R(J)$ of Y, the following sequence is exact:

$$0 \to \mathcal{F}(Y_R(I) \cup Y_R(J)) \to \mathcal{F}(Y_R(I)) \oplus \mathcal{F}(Y_R(J)) \to \mathcal{F}(Y_R(I) \cap Y_R(J)).$$

However, since one easily checks that $Y_R(I) \cap Y_R(J) = \mathcal{K}(\tau_I \vee \tau_J)$ and that $Y_R(I) \cup Y_R(J) = \mathcal{K}(\tau_I \wedge \tau_J)$, this sequence reduces to

$$0 \to Q_{\tau_I \wedge \tau_J}(M) \to Q_{\tau_I}(M) \oplus Q_{\tau_J}(M) \to Q_{\tau_I \vee \tau_J}(M),$$

and the latter is exact by (3.11), since τ_I and τ_J are mutually compatible.

Let us now fix an open subset $Y_R(I)$ and consider the canonical map

$$\theta_I : \varinjlim Q_J(M) \to Q_{\tau_I}(M),$$

the inductive limit being taken over the family of all twosided ideals J of R, with the property that $Y_R(I) \subseteq X_R(J)$. It is easy to see that these morphisms θ_I glue together to a morphism of sheaves $\theta : \mathcal{O}_M|Y \to \mathcal{F}$.

Let $P \in Y$ and let us calculate the induced stalk morphism

$$\theta_P : (\mathcal{O}_M|Y)_P \to \mathcal{F}_P.$$

Let τ_P denote the join of all σ_J, where J runs over the twosided ideals of R with the property that $P \in X_R(J)$ and that $X_R(J)$ belongs to $T(R)$. Then

$$(\mathcal{O}_M|Y)_P = \varinjlim_{P \in Y_R(I)} (\varinjlim_{Y_R(I) \subseteq X_R(J)} \mathcal{O}_M(X_R(J))) =$$
$$= \varinjlim_{P \in X_R(J)} Q_J(M) = Q_{\tau_P}(M),$$

cf. (1.9) in the previous Chapter. On the other hand, denote by τ'_P the join of all τ_I, where I runs through the twosided ideals of R with the property that $P \in Y_R(I)$ and that $Y_R(I)$ belongs to the topology on Y induced by $T(R)$. From the definition of $\mathcal{F}$, it then follows that

$$\mathcal{F}_P = \varinjlim_{P \in Y_R(I)} \mathcal{F}(Y_R(I)) = \varinjlim_{P \in Y_R(I)} Q_{\tau_I}(M) = Q_{\tau'_P}(M).$$

However, as

$$\mathcal{K}(\sigma) = \bigcap \{X_R(J) \in T(R); \ \sigma_J \leq \sigma\},$$

we obtain that $\mathcal{K}(\tau_P) \subseteq \mathcal{K}(\sigma)$, whence

$$\mathcal{K}(\tau_P) = \mathcal{K}(\tau_P) \cap \mathcal{K}(\sigma)$$
$$= \bigcap \{X_R(J) \cap \mathcal{K}(\sigma); \ P \in X_R(J), \ X_R(J) \in T(R)\}$$
$$= \bigcap \{Y_R(I); \ P \in Y_R(J)\}.$$

It thus follows that $\tau_P = \tau'_P$, and from this and the foregoing calculations, one easily deduces that θ_P is actually an isomorphism. As this holds for *any* $P \in Y$, we find that θ is an isomorphism. In particular, this yields that $\theta(Y) : (\mathcal{O}_M|Y)(Y) \to \mathcal{F}(Y) = Q_\sigma(M)$ is an isomorphism, which proves the assertion. $\qquad\square$

The previous proof also yields:

(5.5) Corollary. *For any left R-module M, and any prime ideal P of R, we have* $\mathcal{O}_{M,P} = Q_{(R\backslash P)}(M)$, *where* $Q_{(R\backslash P)}$ *is the localization functor at*

$$\sigma_{(R\backslash P)} = \bigvee \{\sigma_I; \ X_R(I) \in T(R), \ P \in X_R(I)\}.$$

(5.6) Let us now assume that $P \in Spec(R)$ has the property that $\mathcal{L}(\sigma_{R \backslash P})$ possesses a basis of twosided ideals I with $\sigma_I \in \mathcal{S}_{tc}(R)$. Then, obviously, $\sigma_{R \backslash P}$ is a biradical.

On the other hand, in this case, we claim that for any $M \in R-\mathbf{mod}$, we have $\mathcal{O}_{M,P} = Q_{R \backslash P}(M)$. Indeed, let $\{I_\alpha;\ \alpha \in A\}$ be a basis of twosided ideals of $\mathcal{L}(\sigma_{R \backslash P})$ with $\sigma_{I_\alpha} \in \mathcal{S}_{tc}(R)$ for any $\alpha \in A$. Then the corresponding $X_R(I_\alpha) \in T(R)$ form a basis of open neighbourhoods for P, so (as the σ_α are symmetric),

$$\begin{aligned}
\mathcal{O}_{M,P} &= \varinjlim\{Q_I(M);\ X_R(I) \in T(R),\ P \in X_R(I)\} \\
&= \varinjlim\{Q_{I_\alpha}(M);\ \alpha \in A\} \\
&= Q_{R \backslash P}(M),
\end{aligned}$$

which proves the assertion.

In order to calculate stalks in the module-finite case, we will use the following result:

(5.7) Lemma. *Assume the ring R to be module-finite over its noetherian center C. Let $P \in Spec(R)$ and $\mathfrak{p} = P \cap C$. Then*

$$\sigma_{Cl(P)}\left(= \bigwedge_{Q \in Cl(P)} \sigma_{C(Q)}\right) = \overline{\sigma_{C \backslash \mathfrak{p}}},$$

where $\overline{\sigma_{C \backslash \mathfrak{p}}}$ is the radical in $R-\mathbf{mod}$ induced by the radical $\sigma_{C \backslash \mathfrak{p}}$ in $C-\mathbf{mod}$.

Proof. Since the ring R is FBN, all radicals are symmetric. Hence

$$\sigma_{Cl(P)} = \bigwedge_{Q \in Cl(P)} \sigma_{R \backslash Q},$$

so a twosided ideal I belongs to $\mathcal{L}^2(\sigma_{Cl(P)})$ if and only if $I \not\subset Q$ for all $Q \in Cl(P)$.

First, assume $I \in \mathcal{L}^2(\sigma_{Cl(P)})$. Since R is noetherian, the radical of I is a finite intersection $rad(I) = P_1 \cap \ldots \cap P_n$ of prime ideals of R, and there exists some positive integer t with $rad(I)^t \subseteq I$. For each $1 \leq i \leq n$, put $\mathfrak{p}_i = P_i \cap C$. We claim that $\mathfrak{p}_1 \cap \ldots \cap \mathfrak{p}_n \not\subseteq \mathfrak{p}$.

Indeed, otherwise, if $\mathfrak{p}_1 \cap \ldots \cap \mathfrak{p}_n \subseteq \mathfrak{p}$, then $\mathfrak{p}_1 \subseteq \mathfrak{p}$, for example. From (2.4) in Chapter II, it follows that there exists some prime ideal $Q \supseteq P_1$ of R, with $Q \cap C = \mathfrak{p} = P \cap C$. Mueller's result (7.8) in Chapter II then implies that $Q \in Cl(P)$. However, since

$$I \subseteq rad(I) = P_1 \cap \ldots \cap P_n \subseteq P_1 \subseteq Q$$

this leads to a contradiction, which proves our claim.

It thus follows that $(\mathfrak{p}_1 \cap \ldots \cap \mathfrak{p}_n)^t \not\subseteq \mathfrak{p}$, so $I \cap C \not\subseteq \mathfrak{p}$. But then $I \cap C \in \mathcal{L}(\sigma_{C\backslash\mathfrak{p}})$, i.e., $I \in \mathcal{L}^2(\overline{\sigma_{C\backslash\mathfrak{p}}})$.

Conversely, if $I \in \mathcal{L}^2(\overline{\sigma_{C\backslash\mathfrak{p}}})$, then $I \cap C \not\subseteq \mathfrak{p}$. Hence $I \not\subseteq Q$ for all $Q \in Cl(P)$, as for these $Q \cap C = \mathfrak{p}$. So, $I \in \mathcal{L}^2(\sigma_{Cl(P)})$. This proves our assertion. $\qquad \square$

We may now prove:

(5.8) Proposition. *Assume the ring R to be module-finite over its noetherian center C. Let $P \in Spec(R)$ and $\mathfrak{p} = P \cap C$. Then, for any left R-module M, we have $\mathcal{O}_{M,P} = M_{\mathfrak{p}}$.*

Proof. First note that it easily follows from (5.2) and the previous result, that

$$Q_{Cl(P)}(M) = M_{\mathfrak{p}} = \varinjlim_{f \notin \mathfrak{p}} M_f,$$

where $M_{\mathfrak{p}}$ denotes the usual (central) localization of M at the prime ideal $\mathfrak{p}$.

On the other hand, we claim that for any twosided ideal I with $X_R(I) \in T(R)$ and any $Q \in Cl(P)$, we have $\sigma_I \leq \sigma_{R\backslash P}$ if and only if $\sigma_I \leq \sigma_{R\backslash Q}$. Indeed, since the ring R is FBN, it satisfies the strong second layer condition, so from the fact that σ_I is a centralizing biradical, it follows that I satisfies the Artin-Rees property. Since from (3.11) in Chapter III, we then know that for any pair of linked prime ideals $P' \rightsquigarrow P''$ we have $I \subseteq P'$ if and only if $I \subseteq P''$, this proves our claim.

In particular, for any $X_R(I) \in T(R)$ containing P, there is a canonical morphism $Q_I(M) \to Q_{Cl(P)}(M)$, and these glue together to a morphism

$$\varinjlim Q_I(M) \to Q_{Cl(P)}(M) = M_{\mathfrak{p}}.$$

Since the composition

$$\varinjlim_{\substack{f \in C \\ f \notin \mathfrak{p}}} M_f \to \varinjlim_{\substack{X(I) \in T(R) \\ P \in X(I)}} Q_I(M)Q_I(M) = \mathcal{O}_{M,P} \to M_{\mathfrak{p}}$$

is just the identity, clearly $\mathcal{O}_{M,P}$ maps onto $M_{\mathfrak{p}}$.

To conclude, let us show that the canonical map $\mathcal{O}_{M,P} \to M_{\mathfrak{p}}$ is also injective. So, pick $X_R(I) \in T(R)$ not containing P and consider the exact sequence

$$0 \to \sigma_{C \backslash \mathfrak{p}} Q_I(M) \to Q_I(M) \to M_{\mathfrak{p}}$$

If $q \in \sigma_{C \backslash \mathfrak{p}} Q_I(M)$, then $fq = 0$, for some $f \notin \mathfrak{p}$. Let $J = (f)$ and denote by q' the image of q under the morphism $Q_I(M) \to Q_{IJ}(M)$, then $IJq' = 0$, i.e., $q' = 0$, as $Q_{IJ}(M)$ is σ_{IJ}-torsionfree. Since it is also clear that J as well as IJ satisfy the Artin-Rees property, it thus follows that

$$\varinjlim \sigma_{C \backslash \mathfrak{p}} Q_I(M) = \sigma_{C \backslash \mathfrak{p}} \varinjlim Q_I(M) = 0.$$

So the induced map $\varinjlim Q_I(M) \to M_{\mathfrak{p}}$ is injective, indeed. This proves the assertion. $\square$

(5.9) Example. Let us consider the example given in (1.9) in Chapter II. The nonzero prime ideals of the ring

$$R = \begin{pmatrix} D & \mathfrak{m} \\ \mathfrak{m} & D \end{pmatrix}$$

where $D = \mathbf{C}[X]$ and $\mathfrak{m} = (X)$, are the maximal ideals of the form $M_\alpha = (X - \alpha)R$, for $0 \neq \alpha \in \mathbf{C}$ and the two maximal ideals

$$M_+ = \begin{pmatrix} \mathfrak{m}\,\mathfrak{m} \\ \mathfrak{m}\,D \end{pmatrix} \text{ resp. } M_- = \begin{pmatrix} D\,\mathfrak{m} \\ \mathfrak{m}\,\mathfrak{m} \end{pmatrix}.$$

For any $\alpha \in \mathbf{C}$, put $D_\alpha = \mathbf{C}_{(X-\alpha)}$ and denote by $\mathfrak{m}_\alpha$ the (unique) maximal ideal of D_α. Using the previous result, it follows that the stalks of the structure sheaf $\mathcal{O}_R$ are given by

$$\mathcal{O}_{R,M_\alpha} = R_{(X-\alpha)} = M_2(D_\alpha),$$

for $\alpha \neq 0$ and

$$\mathcal{O}_{R,M_+} = \mathcal{O}_{R,M_-} = \begin{pmatrix} D_0\,\mathfrak{m}_0 \\ \mathfrak{m}_0\,D_0 \end{pmatrix}.$$

The ring $\mathcal{O}_{R,M_\alpha}$ has the unique maximal ideal $N_\alpha = M_2(\mathfrak{m}_\alpha)$. Moreover, $M_2(\mathfrak{m}_0) \cap R = M_\alpha$. On the other hand, $\mathcal{O}_{R,M_+} = \mathcal{O}_{R,M_-}$ has maximal ideals

$$N_+ = \begin{pmatrix} \mathfrak{m}_0\,\mathfrak{m}_0 \\ \mathfrak{m}_0\,D_0 \end{pmatrix} \text{ resp. } N_- = \begin{pmatrix} D_0\,\mathfrak{m}_0 \\ \mathfrak{m}_0\,\mathfrak{m}_0 \end{pmatrix},$$

and $N_+ \cap R = M_+$ resp. $N_- \cap R = M_-$. The two maps

$$R \to \mathcal{O}_{R,M_+} = \mathcal{O}_{R,M_-} \to \mathbf{C},$$

induced by the isomorphisms $\mathcal{O}_{R,M_+}/N_+ \cong \mathbf{C}$ resp. $\mathcal{O}_{R,M_-}/N_- \cong \mathbf{C}$ have kernels M_+ resp. M_-. This permits to differentiate the two maximal ideals M_+ and M_- within their common stalk. (See [87] for a similar approach).

(5.10) Let us now take a look at the functorial behaviour of the ringed space $(Spec(R), T(R), \mathcal{O}_R)$. For the definition of centralizing resp. strongly normalizing extensions, we refer to (1.1) resp. (3.2) in Chapter II.

First, consider a centralizing extension $\varphi : R \to S$. If σ is stable or even bistable, then $\varphi_*\sigma = \overline{\sigma}$ does not necessarily share the same property. Fortunately, things are different if we work with centralizing radicals. Indeed, if (σ, ρ) is a centralizing biradical over R and if M is a centralizing S-bimodule, then φ also makes M into a centralizing R-bimodule, so (applying (4.11) in the previous Chapter)

$$\overline{\sigma}M = \sigma M = \rho M = \overline{\rho}M.$$

In particular, if $\sigma_I \in \mathcal{S}_{tc}(R)$, then $\varphi_*\sigma_I = \overline{\sigma_I} = \sigma_{SI} \in \mathcal{S}_{tc}(S)$. As we know that any centralizing extension $\varphi : R \to S$ induces a morphism

$$^a\varphi : Spec(S) \to Spec(R),$$

with the property that $(^a\varphi)^{-1}(X_R(I)) = X_S(SI)$ for any twosided ideal I of R, it thus follows that φ yields a continuous morphism

$$^a\varphi : (Spec(S), T(S)) \to (Spec(R), T(R)).$$

On the other hand, if $\varphi : R \to S$ is a strongly normalizing extension, the image $\varphi_*\sigma$ of a centralizing biradical σ in $R-\mathbf{mod}$ is not necessarily a centralizing biradical in $S-\mathbf{mod}$, due to the fact that a centralizing bimodule over S is, in general, not centralizing over R. However, if M is a centralizing S-bimodule, then clearly M is a strongly normalizing bimodule over R, when endowed with the R-bimodule structure induced through φ.

Now, we have seen in (4.13) in the previous Chapter, that the notions of biradical, centralizing and strongly normalizing biradical coincide, whenever the base ring satisfies the strong second layer condition. Essentially the same proof as in the centralizing case then shows that if R satisfies the strong second layer condition, then the image $\varphi_*\sigma$ of a centralizing biradical σ in $R-\mathbf{mod}$ is a centralizing biradical in $S-\mathbf{mod}$.

So we have proved:

(5.11) Proposition. *Let $\varphi : R \to S$ be a ring homomorphism. Then φ induces a continuous map*

$$^a\varphi : (Spec(S), T(S)) \to (Spec(R), T(R))$$

whenever one of the following statements holds:

(5.11.1) *φ is a centralizing extension;*

(5.11.2) *φ is a strongly normalizing extension and R satisfies the strong second layer condition.*

(5.12) Let us now see what happens at the sheaf level. Recall that we have proved in (6.12) in Chapter II, that if $\varphi : R \to S$ is a strongly normalizing extension between (noetherian) prime rings and if σ is a symmetric radical in $R-\mathbf{mod}$ with induced radical $\overline{\sigma}$ in $S-\mathbf{mod}$, then φ extends (uniquely) to a ring homomorphism $\overline{\varphi}$: $Q_\sigma(R) \to Q_{\overline{\sigma}}(S)$, if and only if $Q_\sigma(Ker(\varphi))$ is a twosided ideal of $Q_\sigma(R)$. Now, it follows from (5.5) or (6.6) in Chapter II, that if σ is ideal invariant, then $Q_\sigma(I)$ is a twosided ideal of $Q_\sigma(R)$ for any twosided ideal I of R.

This proves that if σ is an ideal invariant radical in $R-\mathbf{mod}$ and if $\varphi : R \to S$ is a strongly normalizing extension between rings which are both prime and noetherian, then σ induces a ring homomorphism $\varphi_\sigma : Q_\sigma(R) \to Q_{\overline{\sigma}}(S)$.

Applying this to centralizing radicals of the form σ_I, this yields that if $\sigma_I \in \mathcal{S}_{tc}(R)$ and if $\varphi : R \to S$ is as before, then we obtain a ring homomorphism $\varphi_I : Q_I(R) \to Q_{SI}(S)$, which uniquely extends φ.

Since for any twosided ideal I of R, we have $(^a\varphi)^{-1}(X_R(I)) = X_S(SJ)$, glueing together these morphisms φ_I to a morphism $\varphi_\# : \mathcal{O}_R \to (^a\varphi)_*\mathcal{O}_S$, yields the following extension of (5.11):

(5.13) Theorem. *Let $\varphi : R \to S$ be a ring homomorphism between noetherian prime rings. Then φ induces a morphism of ringed spaces*

$$(^a\varphi, \varphi_\#) : (Spec(S), T(S), \mathcal{O}_S) \to (Spec(R), T(R), \mathcal{O}_R),$$

whenever one of the following statements holds:

(5.13.1) φ *is a centralizing extension;*

(5.13.2) φ *is a strongly normalizing extension and R satisfies the strong second layer condition.*

Note that it is obvious from its definition, that this morphism $(^a\varphi, \varphi_\#)$ of ringed spaces yields back the extension $\varphi : R \to S$, by taking global sections.

(5.14)   Let us now assume that R satisfies the strong second layer condition and that every clique in R is classically localizable. As before, for any link closed subset $X \subseteq Spec(R)$, we define $\sigma_X = \bigwedge\{\sigma_P; \ P \in X\}$. As pointed out in the proof of (7.9) in Chapter II, σ_X is then a biradical, say with associated radical $\rho_X = \bigwedge\{\rho_P; \ P \in X\}$ in $\mathbf{mod}{-}R$, where ρ_P is the right analogue of σ_P.

If $X_R(I) \in T(R)$ for some twosided ideal I of R, then $\mathcal{K}(\sigma_I) = X_R(I)$ is link closed, as σ_I is a biradical, cf. (4.10) in the previous Chapter. Denote by $\sigma_{(I)}$ resp. $\rho_{(I)}$ the radical $\sigma_{X_R(I)}$ resp. $\rho_{X_R(I)}$ associated to $X_R(I)$ as just indicated. Then $(\sigma_{(I)}, \rho_{(I)})$ is a (not necessarily symmetric) biradical, which thus satisfies the Artin-Rees property.

Of course, if R is an FBN ring, then $\sigma_{(I)}$ resp. $\rho_{(I)}$ is just the usual symmetric radical σ_I resp. ρ_I associated to the twosided ideal I of R.

(5.15)   Define a map

$$\gamma : T(R) \to R{-}\mathbf{rad} : X_R(I) \mapsto \sigma_{(I)}.$$

Then it is clear that γ is an order inverting lattice morphism, as both $X_R(I)$ and $\sigma_{(I)}$ only depend upon the radical of I.

On the other hand, obviously γ is continuous, since

$$\bigwedge_\alpha \sigma_{(I_\alpha)} = \bigwedge_\alpha \bigwedge_{P \in X_R(I_\alpha)} \sigma_P = \bigwedge_{P \in \bigcup_\alpha X_R(I_\alpha)} \sigma_P = \sigma_{\bigcup_\alpha X_R(I_\alpha)} = \sigma_{(\sum_\alpha I_\alpha)}.$$

Finally, from (3.25), it follows that for any pair of open sets $X_R(I)$, $X_R(J) \in T(R)$, the radicals $\sigma_{(I)}$ and $\sigma_{(J)}$ are mutually compatible on any normalizing R-bimodule (as they both satisfy the Artin-Rees property, by the previous discussion).

Applying (4.14) thus shows that for any normalizing R-bimodule M, associating $Q_{(I)}(M)$, the localization of M at $\sigma_{(I)}$, to any $X_R(I) \in T(R)$, defines a sheaf $\mathcal{O}'_M = \mathcal{O}^\gamma_M$ on $(Spec(R), T(R))$. This construction applies, in particular, to $M = R$, thus yielding a sheaf of rings $\mathcal{O}'_R$ on $(Spec(R), T(R))$ and any $\mathcal{O}'_M$ thus becomes automatically a sheaf of left $\mathcal{O}'_R$-modules on the ringed space $(Spec(R), T(R), \mathcal{O}'_R)$.

Note that $\mathcal{O}'_R$ also represents R (by taking global sections) and that $\mathcal{O}_M$ and $\mathcal{O}'_M$ coincide for any any normalizing R-bimodule, whenever R is an FBN ring. In general, these sheaves are different, however, one being obtained through symmetric localization and the other through classical localization.

Finally, note that taking sections over arbitrary generically closed subsets in $(Spec(R), T(R))$ yields:

(5.16) Proposition. *Let σ be a symmetric radical in $\mathcal{S}_{tc}(R)$, i.e., let $Y = \mathcal{K}(\sigma)$ be a generically closed subset of $(Spec(R), T(R))$. If M is a left R-module, then*

$$\mathcal{O}'_M(\mathcal{K}(\sigma)) = (\mathcal{O}'_M|Y)(Y) = Q_{\mathcal{K}(\sigma)}(M),$$

where $Q_{\mathcal{K}(\sigma)}(M)$ is the localization of M at the radical $\sigma_{\mathcal{K}(\sigma)}$ in $R-\mathbf{mod}$ associated to the (link closed) subset $\mathcal{K}(\sigma) \subseteq Spec(R)$.

Proof. The proof of this statement may be given along the lines of that of (5.4), except that one has to invoke (1.8) in the previous Chapter instead of (1.9). Indeed, this follows essentially from the fact that our assumptions imply the notions of (centralizing) biradicals and bistable radicals to coincide. We leave details to the reader. $\square$

(5.17) Functoriality is a little more tricky for the latter ringed space, since (6.12) in Chapter II may no longer be invoked (as the radicals we use are not necessarily symmetric).

Let $\varphi : R \to S$ be a ring homomorphism, which we assume to be a centralizing extension, a finite strongly normalizing extension or a strongly normalizing extension, with S (and hence R) prime and where we assume, as before, R to be a (noetherian) ring satisfying the strong second layer condition and with the property that every clique is classically localizable.

Let $X_R(I) \in T(R)$, then $X_R(I)$ is link closed. So it follows from (7.10) in Chapter II, that the left and right localization at $X_R(I)$, i.e., the localization at $\sigma_{(I)}$ resp. $\rho_{(I)}$ of S are isomorphic R-bimodules over S. In particular, it follows that $Q_{(I)}(S)$ is then endowed with a left S-module structure extending that of S. Indeed, the left S-multiplication by $s \in S$ on $Q_{(I)}(S)$ may be defined as the extension to $Q_{(I)}(S) \cong Q_{X_R(I)}(S)$ of the right R-linear map

$$\lambda_s : S \to S : x \mapsto sx.$$

From (5.3) in Chapter II, we derive that $Q_{(I)}(S)$ is a ring, canonically isomorphic with $Q_{\overline{\sigma_I}}(S)$, where $\overline{\sigma_I} = \varphi_* \sigma_I = \varphi_{SI}$. In particular, the extension $Q_{(I)}(R) \to Q_{(I)}(S)$ of φ is actually a ring homomorphism $\varphi_{(I)} : Q_{(I)}(R) \to Q_{(SI)}(S)$ fitting into a commutative diagram

$$
\begin{array}{ccc}
R & \xrightarrow{\ \varphi\ } & S \\
\downarrow & & \downarrow \\
Q_{(I)}(R) & \xrightarrow[\varphi_{(I)}]{} & Q_{(SI)}(S)
\end{array}
$$

As in the symmetric case, these morphisms glue together to a morphism $\varphi'_{\#} : \mathcal{O}'_R \to (^a\varphi)_* \mathcal{O}'_S$, yielding:

(5.18) Theorem. *Let $\varphi : R \to S$ be a ring homomorphism and assume R to be a noetherian ring, which satisfies the strong second layer*

condition and with the property that every clique in $Spec(R)$ is classically localizable. Then φ induces a morphism of ringed spaces

$$({}^a\varphi, \varphi'_{\#}) : (Spec(R), T(S), \mathcal{O}'_S) \to (Spec(R), T(R), \mathcal{O}'_R),$$

whenever one the following statements hold:

(5.18.1) φ is a centralizing extension;

(5.18.2) φ is a finite strongly normalizing extension;

(5.18.3) φ is strongly normalizing and S (and hence R) is prime.

As in the previous, symmetric construction, this morphism $({}^a\varphi, \varphi'_{\#})$ of
ringed spaces permits us to recover the original extension $\varphi : R \to S$
by taking global sections.

Epilogue

At this point, we want to stress the fact that the results obtained
in this Section (i.e., the representations of an arbitrary left R-module
by (structure) sheaves on $Spec(R)$ endowed with suitable topologies),
should be viewed as one of the primary aims and motivations of this
text. The results referred to, show that the machinery we developed
allows to develop an *algebraic geometry* based upon noetherian rings
satisfying the second layer condition and to study these, using geometric and topological techniques.

The reader should compare the present set-up with that developed in
[87], where only rings satisfying a polynomial identity were considered.
Let us point out that the topologies we use are somewhat coarser than
the Zariski topology considered in [87], as we restrict to open subsets
tightly connected to stability or the Artin-Rees property. On the other

hand, the sheaves we construct behave in a much nicer way. In fact, they function outside of the realm of pi rings, they behave functorially with respect to a much wider class of ring extensions and they apply to rings which are no longer assumed to be prime, as well as to modules over them. In this way, they yield a complete solution to most of the unsolved problems left over in [87].

At the end of this text, the reader should realize that we are yet only at the beginning, i.e., that we just *started* the study and applications of the type of noncommutative algebraic geometry we wanted to consider, most of these falling outside of the limited scope of the present text.

Bibliography

[1] Abbas, N.M. and Van Oystaeyen, F., *Algebraically birational extensions*, University of Antwerp, UIA, preprint, 1991.

[2] Anderson, F.W. and Fuller, K.R., Rings and categories of modules, Springer Verlag, Berlin, 1974.

[3] Artin, M., *On Azumaya algebras and finite dimensional representations of rings*, J. Algebra **11** (1969) 532-563.

[4] Artin, M. and Schelter, W., *Integral ring homomorphisms*, Adv. in Math. **39** (1981) 289-329.

[5] Barou, G., Cohomologie locale d'algèbres de Lie nilpotentes, Thèse de 3ième cycle, Université P. et M. Curie, 1972.

[6] Barou, G. and Malliavin, M.P., *Sur la résolution injective minimale de l'algèbre enveloppante d'une algèbre de Lie résoluble*, J. Pure Appl. Algebra **37** (1985) 1-25.

[7] Bartijn, J., Flatness, completions, regular sequences, un ménage à trois, ph. d. thesis, Utrecht, 1985.

[8] Beachy, J., *Stable torsion radicals over F.B.N. rings*, J. Pure Appl. Algebra **24** (1982) 235-244.

[9] Bell, A.D., Notes on localization in noncommutative noetherian rings, Cuadernos de Algebra 9, Universidad de Granada, Granada, 1988.

[10] Bell, A.D., *Localization and ideal theory in Noetherian strongly group-graded rings*, J. Algebra **105** (1987) 76-115.

[11] Bell, A.D., *Localization and ideal theory in iterated differential operator rings*, J. Algebra **106** (1987) 376-402.

[12] Bit-David, J. and Robson, J.C., *Normalizing extensions I*, Lecture Notes in Mathematics 825, Springer Verlag, Berlin, 1980.

[13] Bit-David, J. and Robson, J.C., *Normalizing extensions II*, Lecture Notes in Mathematics 825, Springer Verlag, Berlin, 1980.

[14] Boratynski, M., *A change of rings theorem and the Artin-Rees property*, Proc. Amer. Math. Soc. **53** (1975) 307-310.

[15] Braun, A. and Warfield, R.B., *Symmetry and Localization in Noetherian Prime PI Rings*, J. of Algebra **118** (1988) 322-335.

[16] Braun, A. and Vonessen, N., *Integrality for PI-Rings*, J. Algebra **151** (1992) 39-79.

[17] Brown, K. *The representation theory of noetherian rings*, preprint, University of Glasgow, Department of Math., 1990.

[18] Bueso, J.L., Jara, P. and Verschoren, A., *Extensions and localization*, Comm. in Algebra **19** (1991) 2137-2151.

[19] Bueso, J.L., Segura, M.I. and Verschoren, A., *Strong stability and sheaves*, Comm. in Algebra **19** (1991) 2531-2545.

[20] Bueso, J.L., Torrecillas, B. and Verschoren, A., Local Cohomology and Localization, Pitman Research Notes in Math., New York, 1990.

[21] Bueso, J.L., Jara, P. and Merino, L.M., *A structure sheaf on noetherian ring*, preprint, University of Granada, 1992.

[22] Call, F.W., Torsion theoretic algebraic geometry, Queen's Papers in Pure and Appl. Math. 82, Kingston, Ontario, Canada, 1989.

[23] Chatters, A.W. and Hajarnavis, C.R., Rings with chain conditions, Research Notes in Mathematics 44, Pitman, 1980.

[24] Cauchon, G., *Les T-anneaux, la condition (H) de Gabriel et ses conséquences*, Comm. Algebra **4** (1976) 11-50.

[25] Dixmier, J., Algèbres enveloppantes, Cahiers Scientifiques 37, Gauthier-Villars, Paris, 1974.

[26] Fossum, R., *Invariant Theory, Representation Theory, Commutative Algebra – ménage à trois*, in: Séminaire d'Algebre P. Dubreil et M. P. Malliavin 1980, Lecture Notes in Mathematics vol. 867, Springer Verlag, Berlin, 1981, 1-37.

[27] Gabriel, P., *Des Catégories Abeliennes*, Bull. Soc. Math. France **90** (1962) 323-448.

[28] Galsworthy, J., A sheaf, Heinemann, London, 1917.

[29] Galsworthy, J., Another sheaf, Heinemann, London, 1919.

[30] Godement, R., Théorie des Faisceaux, Hermann, Paris, 1958.

[31] Golan, J., Localization in Noncommutative Rings, M. Dekker, New York, 1975.

[32] Golan, J., Structure sheaves over a noncommutative ring, M. Dekker, New York, 1980.

[33] Golan, J., Torsion Theories, Longman, New York, 1986.

[34] Goldie, A., *The structure of prime rings under ascending chain conditions*, Proc. London Math. Soc. **8** (1958) 586-608.

[35] Goldman, J., *Rings and modules of quotients*, J. Algebra **13** (1969) 10-47.

[36] Goldston, B. and Mewborn, A.C., *A structure sheaf for non-commutative noetherian rings*, J. Algebra **47** (1977) 18-28.

[37] Goodearl, K.R. and Warfield, R.B., An introduction to noncommutative noetherian rings, London Math. Society Student Text Series, Cambridge University Press, London, 1989.

[38] Grothendieck, A. and Dieudonné, J., Elements de Géométrie Algébrique I, Publ. Math. I.H.E.S., 1960.

[39] Hartshorne, R., Algebraic Geometry, Graduate Texts in Mathematics 52, Springer Verlag, 1977.

[40] Heinicke, A.G., *On the ring of quotients at a prime ideal of a right noetherian ring*, Canad. J. Mathematics **24** (1972) 703-712.

[41] Heinicke, A.G. and Robson, J.C., *Normalizing extensions: prime ideals and incomparability*, J. Algebra **72** (1981) 237-268.

[42] Heinicke, A.G. and Robson, J.C., *Normalizing extensions: prime ideals and incomparability II*, J. Algebra **91** (1984) 142-264.

[43] Hendrickx, B. and Verschoren, A., *A note on compatibility and stability*, Comm. in Algebra **17** (1989) 1971-1979.

[44] Jategaonkar, A.V., *Injective modules and classical localization in noetherian rings*, Bull. Amer. Math. Soc. **79** (1973) 152-157.

[45] Jategaonkar, A.V., Localization in noetherian rings, London Math. Soc. Lecture Notes 98, Cambridge Univ. Press, London, 1986.

[46] Krause, G., *On fully left bounded left noetherian rings*, J. Algebra **23** (1972) 88-99.

[47] Lambek, J. and Michler, G., *The torsion theory at a prime ideal of a right noetherian ring*, J. Algebra **25** (1973) 364-389.

[48] Lesieur, L. and Croisot, R., *Extension au cas non-commutatif d'un théorème de Krull et d'un lemme d'Artin-Rees*, J. Reine Angew. Math. **204** (1960) 216-220.

[49] Letzter, E.S., *Prime ideals in finite extensions of noetherian rings*, J. Algebra **135** (1990) 412-439.

[50] Lorenz, M. and Passman, D.S., *Integrality and normalizing extensions of rings*, J. Algebra **61** (1979) 289-297.

[51] Louden, K., *Stable torsion and the spectrum of an FBN ring*, J. Pure Appl. Algebra **17** (1979) 173-180.

[52] McConnell, J. and Robson, J.C., Noncommutative noetherian rings, Wiley Interscience, New York, 1988.

[53] Matsumura, H., Commutative Ring Theory, Cambridge University Press, Cambridge, 1986.

[54] Merino, L.M., *Strongly prime radical and ring extensions*, Proc. First Belgian-Spanish week on Algebra and Geometry, Antwerp, 1988, 100-105.

[55] Merino, L.M., *Compatibilidad de teorías de torsión*, Actas de las XIII Jornadas Hispano-Lusas de Matematicas, Valladolid, 1988.

[56] Merino, L., Radwan, A. and Verschoren, A., *Strongly normalizing modules and sheaves*, Bull. Soc. Math. Belg., **44** (1993) 272-284.

[57] Merino, L.M. and Verschoren, A., *Symmetry and localization*, Bull. Soc. Math. Belg. **43** (1991) 99-112.

[58] Merino, L.M. and Verschoren, A., *Strongly normalizing extensions*, J. Pure Appl. Algebra **92** (1994) 161-172.

[59] Merino, L.M., Localización y extensiones de anillos noetherianos, ph.d. thesis, Granada, 1991.

[60] Mitchell, B., Theory of Categories, Academic Press, New York, 1965.

[61] Mulet, J. and Verschoren, A., *On compatibility II*, Comm. Algebra **20** (1992) 1897-1905.

[62] Mulet, J. and Verschoren, A., *Strong biradicals and sheaves*, Proc. Contact Franco Belge en Algébre (Colmar), Hermann, Paris, 1994 (to appear).

[63] Mulet, J., Cohomologia local y dualidad en anillos de Gorenstein no conmutativos, ph.d. thesis, Valencia, 1992.

[64] Mueller, B., *Localization in noncommutative noetherian rings*, Canad. J. Math. **28** (1976) 600-610.

[65] Mueller, B., *Localization in fully bounded rings*, Pacific J. Math. **67** (1976) 233-245.

[66] Mueller, B., *Twosided localization in Noetherian PI-rings*, J. Algebra **63** (1980) 359-373.

[67] Page, A., *Extensions normalisantes et condition de Goldie*, Comm. Algebra **15** (1987) 1599-1606.

[68] Papp, Z., *On stable noetherian rings*, Trans. Amer. Math. Soc. **213** (1975) 107-114.

[69] Robson, J.C., *Cyclic and faithful objects in quotient categories with applications to Noetherian simple or Asano rings*, in: Noncommutative Ring Theory, Kent State, 1975.

[70] Robson, J.C., *Prime ideals in intermediate extensions*, Proc. London. Math. Soc. **44** (1982) 372-384.

[71] Rosenberg, A.L., *Noncommutative affine semischemes and schemes*, in: Seminar on Supermanifolds, vol. 26, (D. Leites, editor), Reports of the Department of Mathematics, University of Stockholm, 1988.

[72] Rowen, L.H., Polynomial Identities in Ring Theory, Academic Press, New York, 1980.

[73] Segura, M.I., Tarazona, D. and Verschoren, A., *On compatibility*, Comm. Algebra **17** (1989) 677-690.

[74] Sigurdson, G., *Links between prime ideals in differential operator rings*, J. Algebra **102** (1986) 260-283.

[75] Sigurdson, G., *Ideals in universal enveloping algebras of solvable Lie algebras*, Comm. Algebra **15** (1987) 813-826.

[76] Sim, S.K., *Prime ideals and symmetric idempotent kernel functors*, Nanta Math. **9** (1976) 121-124.

[77] Smith, P.F., *The Artin-Rees property*, in: Séminaire d'Algèbre P. Dubreil et M.P. Malliavin 1981, Lecture Notes in Mathematics 924, Springer Verlag, Berlin, 1982, 197-240.

[78] Strooker, J., Introduction to Categories, Homological Algebra and Sheaf Cohomology, Cambridge University Press, Cambridge, 1978.

[79] Soueif, L., *Normalizing extensions and injective modules, essentially bounded normalizing extensions*, Comm. Algebra **15** (1987) 1607-1619.

[80] Stenström, B., Rings of Quotients, Springer Verlag, 1975

[81] Swan, R., The Theory of Sheaves, Chicago University Press, Chicago, 1964.

[82] Tennison, B.R., Sheaf Theory, Cambridge University Press, Cambridge, 1976.

[83] Van Oystaeyen, F., Prime Spectra in Non-commutative Algebra, Lecture Notes in Mathematics 444, Springer Verlag, Berlin, 1975.

[84] Van Oystaeyen, F., Compatibility of kernel functors and localization functors, Bull. Soc. Math. Belg. 28 (1976) 131-137.

[85] Van Oystaeyen, F., Zariski Central Rings, Comm. Algebra 6 (1978) 1923-1959.

[86] Van Oystaeyen, F. and Verschoren, A., Reflectors and Localization, M. Dekker, New York, 1979.

[87] Van Oystaeyen, F. and Verschoren, A., Noncommutative Algebraic Geometry, Lecture Notes in Mathematics 887, Springer Verlag, Berlin, 1981.

[88] Van Oystaeyen, F. and Verschoren, A., Extending coherent and quasicoherent sheaves on generically closed spaces, J. Algebra 89 (1984) 224-236.

[89] Verschoren, A., Relative Invariants of Sheaves, M. Dekker, New York, 1986

[90] Verschoren, A., Compatibility and Stability, Notas de Matematica 3, Murcia, 1990.

[91] Verschoren, A., Functoriality and Extensions, Comm. Algebra, 21 (1993), 4299-4310.

[92] Verschoren, A. and Vidal, C., Relatively noetherian rings, localization and sheaves: part I, K-Theory, (1994), to appear.

[93] Verschoren, A. and Vidal, C., *Relatively noetherian rings, localization and sheaves*: part II, K-Theory, (1994), to appear.

[94] Vidal, C. and Verschoren, A., *Reflecting torsion*, Bull. Soc. Math. Belg. **45** (1993) 281-299.

[95] Vidal, C., Localización y haces en categorías localmente noetherianas, ph.d. thesis, Universidad de Santiago, 1993.

[96] Warfield, R.B., *Noncommutative localized rings*, Séminaire d'Algèbre P. Dubreil et M.P. Malliavin 1985, Lecture Notes in Mathematics 1220, Springer Verlag, Berlin, 1986, 178-200.

[97] Zadeh, L., *Fuzzy sets*, Information and Control **8** (1965) 338-353.

[98] Zamenhof, I., *Esperanto*, 1887.

[99] Zariski, O. and Samuel, P., Commutative Algebra, 2 vol's, van Nostrand, Princeton, 1958/60.

Index

R-bimodule, 86
σ-closed, 27
σ-injective, 27
τ-compatible, 228
$\mathcal{K}(\sigma)$, 44
$\mathcal{L}(\sigma)$, 23
$\mathcal{Z}(\sigma)$, 44

absolutely torsionfree, 105
Artin-Rees property, 159, 166
associated prime, 49
automorphic element, 94
automorphic ring morphism, 94

basis, 22
biradical, 182
biradical topology, 241
bistable radical, 185
bistable topology, 240
bounded filter, 190
bounded ideal, 76

centralizing, 86, 87
centralizing biradical, 188
centralizing extension, 87
centralizing sequence, 162
classical ring of fractions, 4

classically localizable set, 11
clique, 8
closure, 225
co-artinian twosided ideal, 189
compatible, 110, 220
complement, 151
continuous correspondence, 98
continuous lattice morphism, 242
coprimary module, 56
correspondence, 98
cotertiary module, 52

Deligne's formula, 154
direct image of a sheaf, 199
disjunctive radical, 239

essentially closed, 151

filter of finite type, 22
finite codimension, 190
fully left bounded ring, 76

Gabriel filter, 20
generically closed subset, 246
global sections, 195

hereditary torsion theory, 19

ideal invariant, 121
ideal link, 8
idempotent functor, 12
idempotent kernel functor, 12
incomparability property, 72
intersection property, 71
inverse image of a sheaf, 200
invertible ideal, 164

join, 25

link, 8
link closed, 8
localizable prime ideal, 7
localizable set, 10
localizing homomorphism, 3

meet, 25
module of quotients, 29
mutually compatible, 220, 234

normalizing, 94
normalizing bimodule, 94
normalizing extension, 94
normalizing sequence, 162

Ore condition, 3

perfect radical, 38
preradical, 12
presheaf, 195
presheaf of rings, 200
primary submodule, 56
prime module, 49

property (T), 38

radical, 12
radical of finite type, 24
reflector, 29
regular element, 134
right Artin-Rees property, 164
ring of fractions, 3
ringed space, 200

second layer condition, 61
sections, 195
separated presheaf, 197
sheaf, 197
sheaf of rings, 200
sheafification, 198
skew polynomial ring, 103
stable, 20
stable topology, 240
stalk, 199
strong second layer condition, 61
strongly automorphic, 104
strongly normalizing, 102
strongly normalizing biradical, 188
symmetric radical, 41
symmetrization, 113

tertiary decomposition, 54
tertiary module, 52
torsion theory, 15

uniform filter, 225
uniform module, 61

weak Artin-Rees property, 173
weak biradical, 228
weakly regular element, 134

Zariski topology, 89
Zariski-central, 183